RENEWALS 458-4574	DATE DUE		
GAYLORD			PRINTED IN U.S.A.

Fast and Efficient Context-Aware Services

WILEY SERIES IN COMMUNICATIONS NETWORKING & DISTRIBUTED SYSTEMS

Series Editor: David Hutchison, *Lancaster University*
Series Advisers: Harmen van As, *TU Vienna*
                    Serge Fdida, *University of Paris*
                    Joe Sventek, *Agilent Laboratories, Edinburgh*

The 'Wiley Series in Communications Networking & Distributed Systems' is a series of expert-level, technically detailed books covering cutting-edge research and brand new developments in networking, middleware and software technologies for communications and distributed systems. The books will provide timely, accurate and reliable information about the state-of-the-art to researchers and development engineers in the Telecommunications and Computing sectors.

Other titles in the series:

Fast and Efficient Context-Aware Services

Danny Raz, *Technion, Israel*
Arto Juhola, *VTT Information Technology, Finland*
Joan Serrat-Fernandez, *Universitat Politecnica de Catalunya, Spain*
Alex Galis, *University College London, United Kingdom*

John Wiley & Sons, Ltd

Other Wiley Editorial Offices

John Wiley & Sons Inc., 111 River Street, Hoboken, NJ 07030, USA

Jossey-Bass, 989 Market Street, San Francisco, CA 94103-1741, USA

Wiley-VCH Verlag GmbH, Boschstr. 12, D-69469 Weinheim, Germany

John Wiley & Sons Australia Ltd, 42 McDougall Street, Milton, Queensland 4064, Australia

John Wiley & Sons (Asia) Pte Ltd, 2 Clementi Loop #02-01, Jin Xing Distripark, Singapore 129809

John Wiley & Sons Canada Ltd, 22 Worcester Road, Etobicoke, Ontario, Canada M9W 1L1

Library of Congress Cataloging-in-Publication Data

Fast and efficient context-aware services/Danny Raz . . . [et al.].
   p. cm. - - (Wiley series in communications networking & distributed systems)
Includes bibliographical references and index.
   ISBN-13: 978-0-470-01668-8 (cloth : alk. paper)
   ISBN-10: 0-470-01668-X (cloth : alk. paper)
1. Computer interfaces.   2. Computer network architectures. I. Raz, Danny. II. Series.
TK7887.5.F37 2006
006.3--dc22   2006007166

British Library Cataloguing in Publication Data

A catalogue record for this book is available from the British Library

ISBN-13 978-0-470-01668-8
ISBN-10 0-470-01668-X

Typeset in 11/13 pt Times by Thomson Press (India) Limited, New Delhi, India
Printed and bound in Great Britain by Antony Rowe Ltd, Chippenham, Wiltshire
This book is printed on acid-free paper responsibly manufactured from sustainable forestry
in which at least two trees are planted for each one used for paper production.

Contents

Foreword

Computer networks are the essential infrastructure for very many enterprises and their customers. Their principal purpose is to serve the communication needs of their users, whose expectations of the offered level of service are tending to increase as networks become more established. Performance, security and, more recently, high availability are threads of research being explored with the aim of assuring Quality of Service.

Complementary to these important threads is the notion that contextual information can provide another means to improving service quality. A simple example is user-location information which can cause document printing to be routed to the nearest printer without the user having to discover and specify a device (if, of course, this is what the user wants...). Many more examples have become evident with the growth of wireless networks, mobile users and ubiquitous or pervasive computing than with wired networks and tethered users.

The advantages of context-aware services have yet to be realised in two senses; first, people and enterprises are generally not aware of any need; and second, few such systems have been deployed and experienced by users. Awareness will certainly follow once more systems have been built and tried, and experiences reported.

This book reports on advances in the areas of creation, delivery and also the management of services that are context-aware. It derives from a European Union funded research project called CONTEXT in which active and programmable network technologies play an important part. It is a book which, above all, offers a vision of the future rather than an account of deployed solutions, although it does describe one approach to a solution which was built and evaluated as part of the CONEXT project.

It is a book that makes the reader think about possibilities and technical challenges, and comprehensively covers *context* in its various shapes and forms as it applies to humans and their environment, to communication and network devices and their characteristics, and to information paths and flows and their properties.

The implications of this book for network services are of enormous potential interest, and it is with considerable pleasure that I welcome it as an addition to the Wiley Series in Communications Networking & Distributed Systems.

David Hutchison
Lancaster University
April 2006

Abbreviations

Abbreviation	Description
AAA	Authentication Authorization And Access
AAL	Active Applications Layer
AC	Action Consumer
ACAN	Ad Hoc Context Aware Network
ADC	Analog To Digital Converter
ADSL	Asymmetric Digital Subscriber Line
ALAN	Application Layer Active Networking
ANEP	Active Networking Encapsulation Protocol
API	Applications Programming Interface
AN	Active Network
ASCE	Assurance Condition Evaluator
CA	Certificate Authority
CAS	Context-Aware Service
CASP	Context-Aware Active Service Provider
CANP	Context-Aware Active Network Provider
CA-WDS	Context-Aware Wireless Data Service
CC/PP	Composite Capability / Preference Profiles
CCM	Connection Control and Management
CCO	Context Computation Object
CCP	Context Collection Point
CC	Context Client
CDAC	Code Distribution Action Consumer
CECAC	Code Execution Controller Action Consumer
CIS	Contextual Information Service
CIDS	Context Information Distribution System
CIB	Context Information Base
CIDS	Context Information Dissemination System
CIS	Context Information Source
CM	Context Mediator
CPU	Central Processing Unit
CSA	Context Service Adapter
CRL	Certificate Revocation List
DARPA	Defence Advance Research Projects Agency
DDRD	Dynamic Decentralized Resource Discovery
DiffServ	Differentiated Service
DMC	Decision Making Component
DMTF	Distributed Management Task Force

DSCP	Differentiated Services Code Point
EE	Execution Environment
ForCES	Forwarding And Control Element Separation
FE	Forwarding Element
FTP	File Transport Protocol
GUI	Graphical User Interface
GPRS	General Packet Radio Service
HTTP	Hyper Text Transfer Protocol
HCI	Human Computer Interface
ID	Identity
IETF	Internet Engineering Task Force
I/O	Input/Output
IP	Internet Protocol
IPv6	Internet Protocol version 6
IS	Inference System
JVM	Java Virtual Machine
JPEG	Joint Photographic Experts Group
LAN	Local Area Network
MSISDN	Mobile Station Integrated Services Digital Network
MAC	Media Access Control
MIB	Management Information Base
NodeOS	Node Operating System
NE	Network Element
NPN	New Public Network
OGSA	Open Grid Service Architecture For Distributed Systems Integration
PACL	Permitted Access Control List
PAD	Personal Digital Assistant
PBMS	Policy-Based Management System
PCCS	Perceptual Context Classifier System
PCE	Perceptual Context Engine
PKI	Public Key Infrastructure
PM	Policy Manager
P2P	Peer-To-Peer
SDD	Service Definition Document
SDF	Service Deployment Framework
SICE	Service Invocation Condition Evaluator
SIP	Session Initiation Protocol
SLO	Service Logic Object
SL	Service Layer
SLA	Service Level Agreement
SNMP	Simple Network Management Protocol
SOAP	Simple Object Access Protocol
QoS	Quality of Service
QoC	Quality of Context
UDDI	Universal Description, Discovery And Integration
UDP	User Datagram Protocol

URL	Uniform Resource Locator
UCL	URL Class Loader
VLAN	Virtual Local Area Network
VoIP	Voice Over Internet Protocol
VPN	Virtual Private Network
VoD	Video On Demand
TINA	Telecommunication Information Networking Architecture
TCP	Transport Protocol
WFA	Work From Anywhere
WiFi	Wireless Fidelity
WIMAX	IEEE 802.16 Standard
WLAN	Wireless Local Area Network
WWW	World Wide Web
XSP	Extensible Service Protocol
XML	Extensible Markup Language
XSL	Extensible Stylesheet Language

1

Introduction

1.1. Context-Aware Services

In personal communications, people often condense their speech by omitting information that can be directly deduced from the circumstances: such an awareness of surroundings, or context, assists the efficient exchange of ideas. Further, one of the parties to a discussion may notice or introduce a change in context, and react to this change as the situation demands. As for the nature of this information, in classical rhetorical theory the constituents of *circumstantiae* usually include time, place, events, manner, causes, persons and instruments related to an incident.

In the same way, computer applications could be made responsive to users' wishes if they were context aware, that is capable of inferring the users' true intentions by taking into account any relevant auxiliary information supplied for the purpose. Thus changes in different types of context information could cause a variety of actions to be initiated by the applications, just as a person might respond to the same signals.

This book describes and discusses the underlying principles of a contextware system that can handle the authoring, creation, management and operation of on-the-fly context-aware services, or indeed, any advanced network services, although context-aware ones present the most intriguing challenge. The reader of this book should be well versed in the ways of the Internet, since familiarity with its basic technologies is a prerequisite for embracing the presented ideas. However, although the book is not intended to be a tutorial on the key topics, it will contain reviews of technologies such as Active networks and Management Systems and as such it will gradually introduce the reader to the main subjects of the book. Thus, the book can also be used as a general introduction to the area of advanced telecommunications services for management and support personnel within network and service provider organisations, as well as a detailed reference book for professional technical staff and graduate-level students.

Fast and Efficient Context-Aware Services   Danny Raz, Arto Tapani Juhola,
Joan Serrat-Fernandez, Alex Galis   © 2006 John Wiley & Sons, Ltd

The following paragraphs present the authors' view of context-aware services and the role of selected technologies in the overall solution. This is followed by a preview of the individual chapters of the book.

The notion of networked applications receiving and making use of information about objects and circumstances around them, and thus presenting a context-aware service to users, has received a lot of attention, exemplified by services like location-aware tourist guides available in mobile terminals. Still, although there have been many context-aware systems and applications tested over the last decade, most of them are still prototypes only available in research labs and in academia. One of the main drawbacks lies in the complexity of capturing, representing and processing the contextual data. The implementations have also lacked generality and flexibility in the sense that only a predefined set of context information has been utilised, with no allowance for customisation or augmenting the scope of the information as the need arises. Yet the range of potentially useful context information is limitless and unforeseeable. One notable source of useful pieces of 'raw' context information has been recognised, though: the network.

Until recently, the sole purpose of the Internet infrastructure, that is interconnected routers, was thought to be to move traffic around as efficiently as possible. This was also the state of the affairs when the notion of active, and subsequently programmable networks arrived. The main idea with these new technologies is to allow a network's behaviour to be modified instantaneously and without service interruption.

1.2. The Context Project

On the context-aware service side of things, what the existing systems are missing is malleable and extensible context information processing, in a word, programmability. As it happens, this is the hallmark of active and programmable networks, specifically these networks are able to modify their behaviour on-the-fly. Recalling that the required information in many cases depends on data extracted from the networks, the inescapable conclusion is that context information can only be generated in a flexible manner by active or programmable networks. This foundational idea, presented by Prof. Alex Galis, was the basis for a European Community-funded research project, CONTEXT.

This project implemented and demonstrated an efficient solution for the automated creation, delivery and management of context-aware services using a very practical form of programmable network technology. The degrees of freedom made possible by this approach are notable: The collection and distribution of necessary context information for a service can be arranged by the service designer in parallel with the creation of other service logic.

At the time of writing this book, project CONTEXT has delivered its final reports, which include the conclusion that active/programmable network nodes can and should be augmented with the means to provide network configuration, status and other useful information, to be refined into context information according to an information model specified at the same time as the service needing the information (or later). This has a noteworthy consequence: No standardisation needs to precede the deployment of such models.

Such network context awareness is a potent and unifying ingredient to be added to the arsenal of service developers. Combined with prediction, information passing, proactiveness and other forms of intelligence, context-aware services can offer tangible benefits.

For the further benefit of service developers, a trial service management system was developed, encompassing authoring of the required information models and policies, service creation and deployment subsystems and policy-based management functionality. These project results are applicable to any advanced service making use of programmable network technology, not just the context-aware ones.

1.3. Structure of the Book

A brief summary of the remaining chapters of this book is provided below:

Chapter 2, 'Context-Awareness and Modelling: Background', sets the scene by laying out the principles involved with the expansive modelling of the context information. The chapter also is a short analysis of the current state of the art in Context-Aware Services.

Chapter 3, 'The Service Lifecycle Functional Architecture', shows what steps needs to be taken during the life of a service, and outlines the required functional abstractions to cater for them.

Chapter 4, 'CAS and the Network Layer', describes what is expected from a network to support context-aware services. The chapter also presents the design approach chosen by the authors.

Chapter 5, 'Baseline Technology', presents the starting point that was available for the creators of CONTEXT system. The major technological inputs are highlighted, including programmable network implementations.

Chapter 6, 'CAS Creation and Management – System Architecture and Design Considerations', lays out the fruits of the work carried out to outline and design a concrete system to handle the functional and nonfunctional demands presented.

Chapter 7, 'Active Application Layer – System Architecture and Design Considerations' brings us to the arena of real action, the network proper. The chapter reveals the main characteristics of the Active Application Layer, design approaches of special interest and the fine points of select aspects of active technology.

Chapter 8, 'System Evaluation', describes the methods of evaluation, evaluation criteria, execution of the tests and the results.

Chapter 9, 'Conclusions', pulls together the themes presented in the previous chapters and discusses the possible ways to improve the present system for wider applicability.

1.4. Acknowledgements

This book is a joint effort of the people who were active in project CONTEXT, and the contained ideas and texts are largely drawn from the material created in this project. The editors wish to specially thank the following for their contribution to the CONTEXT project: Dr Panos Georgatsos; Mr Takis Damilatis; Dr Dimitrios Giannakopoulos from ALGONET S.A., Greece; Mr Juan Manuel Sánchez; Mr José Fabian Roa Buendía from Telefónica Investigación y Desarrollo S.A. Unipersonal Spain; Mr Ricardo Marín-Vinuesa; Mr Javier Justo-Castaño; Mr Marín Serrano-Orozco from Universitat Politècnica de Catalunya, Spain; Mr Rami Cohen from TECHNION Israel Institute of Technology; Mr Kimmo Ahola; Ms Titta Ahola from VTT Technical Research Centre of Finland, Information Technology, Finland; Mr Kerry Jean; Mr Nikolaos Vardalachos; Dr Kun Yang[1]; Prof. Chris Todd from University College London, United Kingdom; Ms Irene Sygkouna; Ms Maria Chantzara from Institute of Communication and Computer Systems, National Technical University of Athens, Greece; Mr Takis Papadakis from VODAFONE-PANAFON Hellenic Telecommunications Company S.A., Greece.

We thank Mr Richard Lewis for his helpful comments on improving the readability of the book.

Finally, we would like to thank Mr Jose Fernandez-Villacanas, European Union project Officer, Dr Anxo Cereijo Roibas and Mr Toon Norp, project reviewers, for their support, wisdom and encouragement for the work of the CONTEXT project. They modulated the evolution of the project and therefore favourably affected the content of this book.

The information and the source codes for many system components produced during the project are available at http://context.upc.es/, under a special 'CONTEXT' breed of an open source licence.

12th February 2006

Danny Raz
Arto Tapani Juhola
Joan Serrat Fernandes
Alex Galis

---

[1]Currently at the University of Essex.

2

Context Awareness and Modeling: Background

The purpose of this chapter is to introduce the concepts of Context and Context-aware services, and to present to the reader the current state of the art and the main results relevant to the Context definition, and to Context-aware research.

2.1. Some Context Definitions

Any kind of activity, including the communication between humans, is surrounded and influenced by context. In the same way that a hand gesture or a word has different meanings depending on the situation in which they are expressed, the users of any IT system are also surrounded by their context when they interact with applications or services. As such, context can be a reflection of real or physical characteristics, as well as characteristics of the virtual world that determine the performance of the application and/or service.

Context awareness, as a process, system, and concept, is based on a group of interrelated areas of research: mobile computing, ubiquitous and pervasive computing, ambient computing, serviceware networking, programmable networks, autonomic communications and ambient computing, and a grid computing and networking. In each of these research areas context has been used to enhance human–computer and computer–computer interaction, thereby providing seamless computing and networking anywhere, anytime.

The Merriam-Webster dictionary defines contexts as 'the interrelated conditions in which something exists or occurs.' In our case the 'something' is a service, but the lack of a specific formal definition of Context with respect to services opens the door for innovation and imagination and the term is used in various meanings in different fields of computing and networking.

Fast and Efficient Context-Aware Services   Danny Raz, Arto Tapani Juhola,
Joan Serrat-Fernandez, Alex Galis   © 2006 John Wiley & Sons, Ltd

Dey's review [19,20] provides the following key definition: 'context is any information that can be used to characterize the situation of an entity. An entity is an object, place or person that is considered relevant to the interaction between a user and an application, including the user and applications themselves.' This definition does not cover all aspects of context as it only presents an external behavior viewpoint, revealed in 'characterizing the situation of an entity.' An internal state viewpoint would need to be added to this definition, identifying the structure of context, its domain, range, qualities, functionality, and control.

Context can also be seen as everything around (and possibly within) an entity, including the entity and its interactions [76]. In fact, if a piece of information or knowledge can be used to characterize the situation of an entity in an interaction then it can be identified as a context characteristic. In many cases, researchers use a definition of context that is appropriate from their point of view and interest. Sometimes the definition is very general.

Schilit and Theimer [81] refer to context as location, identities of nearby people and objects, and changes to those objects.

Brown *et al.* [4] identifies context as the elements of the user's environment that the user's computer knows about. As such context is defined as location [35,68,73,74], identities of the people around the user, the time of day, season, temperature, etc.

Schmidt et al. [82,83] identifies context as the knowledge about the user's and device's state, including surroundings and situation.

These general definitions are difficult to apply or use in a larger scale system.

Other definitions of context are, on the other hand, too specific. Dey [19,20] exemplifies context as the user's emotional state, focus of attention, location and orientation, date and time, objects, and people in the user's environment, while Pascoe [69,70] defines context as the subset of physical and conceptual states of interest to a particular entity.

Important aspects of context are identified by Schilit *et al.* in Reference [75]. They are: where you are, who you are with, and what resources are nearby. Examples of types of context are summarized in Reference [19] as location, identity, activity, and time.

In this book, we want to capture the full meaning of Context with respect to telecommunication and data services and to classify the relevant context information. This, as explained above, is not an easy task due to its extreme heterogeneity of Context and the diversity of services. Some examples of characteristics of context information are:

- *Context types*:
  - Human User context characteristics include information representing the user's surroundings (user location, identity, user mobility, available devices, etc.) as well as his/her physical being (e.g., identity, preferences, history, etc.).

- Device context characteristics include [1]: IP address per machine, IP masks per subnetwork or address per domain – parameters that vary according to our preferred level of abstraction [30]. The complexity escalates when we look at the proliferation of mobile devices [41], for example mobile phones and PDAs that now have access to the Internet. In terms of location [55] as context data, the mobile telephony research community has long developed a reliable system for hand-over between base stations and international roaming.
- Network context characteristics: network identity; network resources: bandwidth, available media ports; other parameters: available quality of service (QoS), security level, access-types, coverage. The network Context Information Base (CIB) is a logical construct representing a distributed repository for network context data and operands, and it can be used by all networking functions and services. The CIB's functionality includes: (i) methods and functions for keeping track of context sources, including context registration and naming, context data directory, indexing, context data monitoring and management, etc.; (ii) collection and distribution of context data to clients through context associations, including context data update and context processing such as aggregation, inference, etc. to support higher level context services.
- Flow context characteristics: flows are the physical and electronic embodiment of the interaction between the user and networks. Context information that characterizes these flows may be used to optimize or enhance this interaction including: the state of the links and nodes that transported the flow, such as congestion level, latency/jitter/loss/error rate, media characteristics, reliability, security; the capabilities of the end-devices; the activities, intentions, preferences or identities of the users; or the nature and state of the end-applications that produce or consume the flow. Because of the ephemeral nature of flows, flow context has to be handled differently than user or network context.

- *Persistence*:
  - Permanent (no updating needed): context, which does not evolve in time, remaining constant for the length of its existence (e.g., name, ID card).
  - Temporary (needs updating): part of the context information that changes (e.g., position, health, router interface load).

- *Evolution*: (of temporary context)
  - Static: in this category we can put the context that does not change very quickly. For example, the temperature throughout the day.
  - Dynamic: in this category we can find the context that changes quickly. For example the position of a person who is driving a car.

- *Medium*:
  - Physical (measurable): refers to context information that is tangible, for example geographical position, network resources, temperature, humidity (it

is likely that this kind of information will be measured by sensors spread all over the network).

– Intangible (nonmeasurable by means of physical magnitudes): the remaining context information, for example name, hobbies (it is likely that this kind of information will be introduced by the user or customer themselves).

- *Relevance to a service or application*:
  – Necessary: part of the context information that must be retrieved for a specific service to run properly.
  – Accessory: additional context information which, although not necessary, could be useful for the purpose of providing a better or more complete service.

- *Temporal situation:*
  – Past: This category comprises context information from the past. For example an appointment for yesterday. This category could be considered to be the context history, which contains all previous user contexts. A context trace is a subset of the context history. A context trace will only contain the contexts that are relevant to the situation under consideration.
  – Present: this category is for the current context: where am I at this moment, etc.
  – Future: in this category we find context detailing scheduled or predicted future events. For example, the venue of tomorrow's meeting. The future context can include user contexts that either the systems or the users can predict and describe, for example activities in your planner. Prediction of future user context would be useful when the user changes location or when subscribing to a new service.

- *Interaction features*: interaction between Context Sources and Context Sinks can be characterized as follows:
  – Context Push: the context sources periodically push updated context information to the context sinks. The context sinks maintain the information in a context store, from which they service client inquiries.
  – Context Pull: the context sinks must explicitly request context information. They can make these requests either on a periodic basis (polling) or when an application demand arises. Each mechanism has advantages and disadvantages. A polling system collects data ahead of need and thus may offer better performance. However, it may consume substantial resources transferring [52] and storing information that is never required, although this may be worthwhile if information freshness is important. In some circumstances, it may be possible to use prefetch and/or caching mechanisms to alleviate these problems, but this may increase resource utilization.

The examples above describe how context information is gathered and how it evolves over time. However, this discussion gives rise to some questions about how context information should be managed, stored, aggregated, disseminated, and used, considering its changing nature. For example, we might consider it helpful to associate a timestamp, a period of validity, and a Quality of Context [10] with each piece of context information.

2.2. Context-Aware Service

Context-aware computing is a computing paradigm in which applications and services can take advantage of contextual information such as user and device location, state, time of day, nearby places, people and devices, and system and user interactions and activities. Many researchers have explored context-aware computing and developed a number of context-aware services to demonstrate and validate the usefulness, the flexibility, and the service personalization of this new technology. Context-aware system infrastructures [33,34] to support multiple applications and services are extremely difficult to maintain over time. This is due to the lack of standard methods to represent context, use context, and build context-aware services and applications. It is also due to the diverse nature of context and the systems that capture, store, and disseminate context.

A definition of context awareness is given [4] as: a system is context aware if it uses context to provide relevant information and/or services to the user, where relevancy depends on the user's task.

Tuulari [85] proposes two categories of context awareness: self-contained context awareness and infrastructure-based context awareness. The former implies context awareness achieved without any outside support, and the latter implies context awareness achieved with outside support.

Chen and Kotz [4] extended Tuulari's division into four context categories in order to achieve a better understanding of the concept: (i) computing context includes network connectivity, bandwidth, communication costs, and nearby resources such as printers, displays, and workstations; (ii) user context includes user profiles, user location, user mobility and nearby users and people; (iii) physical context includes lighting, temperature, and humidity; and (iv) time context includes time of the day, week, year, and also the season of the year.

Three features for context-aware applications are listed in Reference [19] as follows: presentation of information and services to a user [42], automatic execution of a service, and tagging of context to information for later retrieval. The exploitation of local resources and resource discovery are not explicitly mentioned as being context-aware features in Reference [19] because they are considered to be included in the three features mentioned in the previous sentence.

There are two types of context-aware computing:

- *Using context*: Dey [19,20] defines context awareness to be a work leading to the automation of a software system based on knowledge of the user's context. Pascoe *et al.* [69,70] define context-aware computing as the ability of computing devices to detect and sense, interpret, and respond to aspects of a user's local environment and the computing devices themselves. Salber *et al.* [31,78] define context awareness as the ability to provide maximum flexibility of a computational service based on real-time sensing of context.

• *Adapting to context*: [4,11,12,14,15,17,32,59,67,75,91–95] define context-aware applications to be applications that dynamically change or adapt their behavior based on the context of the application and the user [6–9]. Brown *et al.* [4] define context-aware applications as applications that automatically provide information and/or take actions according to the user's present context as detected by sensors. Environment-directed applications are applications that monitor changes in the environment and adapt their operation [57] according to predefined or user-defined guidelines.

2.3. Context-Awareness System Research

The following section provides a brief review of some of main results in context-awareness system research.

2.3.1. Context-Aware Ubiquitous Computing Applications

There is an increasing interest in computer applications that are aware of the user's context. Currently these applications are normally handcrafted [53,56,58,90]. Many of them present information to users as they enter a given context [54], for example a tourist nearing a site within a city or a visitor moving round a building.

Orr and Abowd [62] describe a system for identifying people based on their footstep ground reaction force (GRF) profiles and how its accuracy was tested against a large pool of footstep data. This floor system may be used to identify users transparently in their everyday living and working environments. They created user footstep models based on footstep profile features and achieved a recognition rate of 93%. They mention that they planned to integrate the system into the Context Toolkit for live use. The Context Toolkit aims to ease the development of context-aware applications by providing a library of 'context widgets' that free the application writer from the details of context sensing (i.e., interfacing with sensors). In the same way that GUI widgets insulate applications from certain interface presentation concerns, context widgets insulate applications from context acquisition concerns. The system consists of these context widgets and a distributed infrastructure that hosts and coordinates the widgets. In order to integrate the Smart Floor with the Context Toolkit, a software layer would have to output the calculated identity of the user (or perhaps the top three choices, along with a certainty score), providing a ready-to-use identity widget, similar to a widget that uses another identification technology such as face recognition or RFID tags. Application writers could then easily use this widget as their interface to the Smart Floor system, without concerning themselves with the details of interfacing to the floor system or with changes to the system as it evolves.

Oppermann and Specht [61] describe the goal and practice of an exhibition guide called Hippie, which takes into account the context of nomadic users. For the context of use three different models are identified: the domain model that describes and classifies the objects of the domain information that are to be presented and processed; the space model that describes the physical space where the nomadic system is used and the location of the domain objects in the space; and the user model that describes the knowledge, the interests, the movement, and the personal preferences of the user.

The main beneficial features of the system for the users are:

(i) Permanent system accessibility: at home the user can access the system using a desktop computer with a high-resolution screen in order to study the site of interest, while on a visit to the exhibition the user takes a handheld computer (PDA) with wireless LAN connection.

(ii) Location awareness allows the system to present information that is relevant to the visitor's current position, identified in two ways. Whether the user is at home or at the exhibition is identified by the type of the device being used. At the exhibition the visitor's location is identified by the infrared infrastructure sensing his position, and the direction he is taking by an electronic compass. These values are transmitted from the handheld computer to the server so that it can automatically send appropriate information to the visitor about the nearest exhibit. The infrared infrastructure consists of emitters installed on the walls underneath each exhibit.

(iii) Multimodal information presentation, which exploits the range of human perception. The information presented during the visit is multimodal, containing written text on the screen and spoken language via headphones. While at the exhibition, the visitors visual attention is free to experience the physical environment, especially for the exhibits. At home, after the visit to the exhibition, additional multimodal information about the exhibition is available including text, graphics, and animations.

(iv) Adaptation to the user's knowledge and interests. The amount of knowledge and level of interest the various users have in the subject in question can vary significantly. The adaptive component runs a user model describing the users' knowledge and the interests. This model automatically evaluates each user's interaction with the exhibition information system and his navigation through the exhibits, that is in the physical space. Importantly, the exhibition system cannot acquire knowledge from external sources, but is restricted to basing its responses on the user's interactions with the system. Alternatively the system can allow the user to specify prominent interests in a user profile dialogue.

Gray and Salber [31] presents a method for analyzing and formulating sensed context information that assists the generation, documentation, and assessment of context-aware application designs. A model of sensed context is analyzed. Sensed context is defined the context that comes from the physical environment, that is that part of context that is accessible via sensors. The sensed context consists of information content and meta-information content. Information content includes the sensed properties of the phenomena (e.g., location, time, identity) and the subjects of the sensing (e.g. location of the person that holds the GPS receiver). The meta-information includes information quality attributes and information about the source of the content. The meta-information's attributes include: forms of representation (e.g. geographical coordinates or a building name for location information), information quality (coverage, resolution, accuracy, sample rate etc.), sensory source, data transformation, and actuation (e.g. shutdown a faulty sensor). The article gives an example of a context-aware museum tour guide whose goal is to deliver information about exhibits in the language of each visitor. The paper analyzes the information and meta-information content in this example. The required information is the visitor's language, the exhibit the visitor is examining, and the exhibit's description. The first two are sensed context. The paper presents the quality criteria of these context examples.

Schilit *et al.* [75] describe systems that examine and react to an individual's changing context. The investigation starts with the definition of 'mobile distributed computing system' and 'context-aware systems.' Four categories of context-aware applications are described: proximate selection, automatic contextual reconfiguration, contextual information and commands, and context-triggered actions. Instances of these application types have been prototyped on the PARCTAB, a wireless palm-sized computer. The aforementioned categories are the product of two points along two orthogonal dimensions: whether the task at hand is getting information or carrying out a command, and whether it is effected manually or automatically. Proximate selection (get information, effected manually) involves a user interface technique where discovered objects that are nearby are emphasized or otherwise made easier to choose. The discovered objects could be input and output devices that require physical interaction, for example printers or displays, non-physical objects and services that are routinely accessed from particular locations, for example bank accounts or menus, or places about which one wants information, for example restaurants or stores. In the above cases, location information can be used to weight the available choices. Automatic Contextual Reconfiguration (get information, effected automatically): Reconfiguration is the process of adding new components, removing existing ones, or altering the connections between components. In the case of context-aware systems, the interesting aspect is how context of use might bring about different system configurations and what these adaptations are. In the case that the context of use is location for example, one possible application is related to the use of the PARCTAB whiteboard, a multi-user drawing program. Entering a room causes an automatic binding between the mobile host and the room's virtual whiteboard, while moving to a different room brings up a different drawing surface. Reconfiguration could be based on other information, in addition to location, such as the people present in a room. For example, if a project group is meeting then the project whiteboard is active. Finally, contextual reconfiguration might also include operating system functions, for example an operating system can use the memory of nearby idle computers for backing store, rather than swapping to a local or remote disk. In this last case, the context of use refers to the hosts in the vicinity.

Want *et al.* [86] present the Active Badge location system, an early solution to the problem of determining location information. This device can help a receptionist locate employees without a public-address system or without telephoning all the possible locations at which they might be found. These kinds of solution can help avoid a great deal of irritation and disruption in offices. Further advantage can be gained from location information by allowing PBX users to define rules governing when a call transfer is allowed. Where you are and who you are with can be used to affect decisions, for instance most people would prefer not to take unexpected telephone calls when they have just been called into their boss's office. Even though this is a relatively old paper, the Active Badge has a historical interest, as it was used as one of the first 'ubicomp' devices at Xerox PARC.

Want and Schillit [87–89] deal with location-aware computer applications that sense their location and modify their settings, user interface, and functions accordingly. The authors perform a brief retrospection over the past few years to verify the notion of increasing computer mobility and ubiquitous connection through wireless networks. With mobility and location in mind, they initiated two research projects: the Active Badge project at Olivetti Research and, later, the PARCTab project at Xerox Palo Alto Research Center. In the case of Active Badge, the person carried an electronic badge, rather than a computer, that informed the computer infrastructure where he was. The infrastructure in turn used his location data to modify the behavior of programs running on nearby workstations. The initial intention was that the Active Badge simply routed telephone calls arriving at the office PBX to the telephone extension nearest the intended recipient. Afterwards it was found that the system could differentiate between a button press and the normal operation of the badge, which meant that a test button could also be used to send commands to the system. These commands could have a personalized meaning for each user and could be interpreted differently in each location. The PARCTab used a true palm-sized tablet computer with a pen interface linked to a diffuse microcellular infrared network. Limited by the technology of the time, the tab commanded the applications, but they were executed in the computer infrastructure while the results were displayed on the screen of a tab, transmitted over the diffuse IP network. The microcellular property provided location information on a room-by-room basis.

2.3.2. Context-Aware Frameworks

Ektara [18,75]. EKTARA defines a full framework for Context-Aware Wearable and Ubiquitous Computing Applications. The EKTARA framework consists of the following components:

- Context-Aware Integration Manager (CAIM). This provides a uniform framework for interaction between applications and the user. The primary goal of the CAIM is to minimize the demands on the user's time and attention while maximizing the relevance of the information provided. The CAIM aims to achieve this goal by taking into account the HCI resources currently available to the user, important contextual factors making the user's ability to pay attention to the UWC system, and the actions of the user's applications. Further, the criteria by which the CAIM makes these decisions must be understandable and controllable by the user. Ideally the CAIM should implicitly learn the user's preferences over time yet still be able to provide the user with an explicit description of its decision model, which becomes part of the user's personal profile, which is always accessible wherever she goes (e.g., by being stored on her person in a wearable computer).
- Contextual Information Service (CIS). This is a distributed database service, which provides UWC applications and services a uniform means of storing and retrieving contextual information. Clients may query a server for all records matching a context template or subscribe to receive records when matching information is posted or expires. CIS Servers may register selected contexts of other servers, allowing clients to discover other members of the CIS federation. The CIS must support a range of context classifications, including location, authorship, intended recipient of information, time of posting, time of relevance, time of expiry, deliverability (whether a record is ordinarily intended to be delivered once or multiple times), MIME document type, and an extensible mechanism to allow the uniform handling of unexpected or idiosyncratic contexts.
- Perceptual Context Engine (PCE). This is a means of turning raw sensor data and information from other sources, such as a real-world description such as 'Meeting in Espoo,' into symbolic context descriptions. The PCE has a two-layer structure, with an inference system running over a perceptual context classifier system.
- The Perceptual Context Classifier System (PCCS) is a signal-processing system, which converts the raw sensor data into a collection of probabilistic estimates. Conceptually, the classifier system allows the user to train event recognition functions, or classifiers, to recognize patterns in the sensory data and tag them as specific events. The mechanism by which this time-series recognition occurs is a multi-level HMM grammar that is capable of recognizing patterns over a range of timescales from seconds to days.
- Inference System (IS). The job of the Inference System is to take the output of the classification system (and other sources of context such as a system clock or GPS

receiver) and convert this information into symbolic context descriptions. The inference system also allows for multiple interpretations of the underlying data, such as continuous (e.g., latitude and longitude) as well as nominal or discrete (e. g., Wordsworth, Garibaldi Square) representations of location.

- Dynamic Decentralized Resource Discovery (DDRD). The dynamic decentralized resource discovery framework allows UWC applications and services to find and use resources that match semantic descriptions of functionality and context. The foundation of this system is a protocol by which a UWC component obtains networking services and contacts a directory registration service. The UWC component provides the registration service a semantic description of itself and its capabilities, and any additional contextual information it chooses to provide. The registration service then makes further determinations about the resource's contextual information server (CIS). If this resource becomes unavailable, the registration service informs the CIS and the registration information is removed.

Dey [23,24] describe Cyberdesk as an architecture that was built to automatically integrate web-based services based on virtual context, or context derived from the electronic world. The virtual context was the personal information the user was interacting with on-screen including email addresses, mailing addresses, dates, names, URLs, etc. An example application is when a user is reading her e-mail and sees that there is an interesting message about some relevant research. She highlights the researcher's name, spurring the CyberDesk architecture into action. The architecture attempts to convert the selected text into useful pieces of information. It is able to see the text as simple text, a person's name, and an email address. It obtains the last piece of information by automatically running a web-based service that convert names to email addresses. With this information, it offers the user a number of services including searching for the text using a web-based search engine, looking up the name in her contact manager, and looking up a relevant phone number using the web. The architecture of Cyberdesk is based on an event-driven model, where components act as event sources and/or event sinks. The system consists of five core components: the Locator, the IntelliButton, the ActOn Button Bar, the desktop and network services, and the type converters. The Locator component in CyberDesk keeps a directory of all the other components in the system, what events they can generate, and/or what events they can consume. The IntelliButton component is the core of the CyberDesk system, as it provides the automatic integrating behavior. It uses the Locator to keep track of all the desktop and network services and the type converters, and all the event sources and sinks they provide. When new components are added to the system, the IntelliButton notifies them that it is interested in all the events that they can generate. The ActOn Button Bar is the user interface for the integrating IntelliButton. The fourth type of component, desktop and network services are the actual services the user wants to access. Desktop services include email browsers, contact managers, and schedulers. Network services include web search engines, telephone directories, and map retrieval tools. Data typing is used extensively in the interface declarations of the event sources and sinks that applications provide. The property field that corresponds to each interface declares the datatype/event that a component is interested in or can provide. The CyberDesk system takes advantage of the Java-type system to do the data typing.

While Cyberdesk was limited in the types of context it could handle, it contained many of the mechanisms that are necessary for a general context-aware architecture. Applications simply specified what context types they were interested in, and were notified when those context types were available. The modular architecture supported automatic interpretation, that is, automatically interpreting individual and multiple pieces of context to produce an entirely new set of derived context.

Salber and Dey [78] introduce the concept of context widgets that mediate between the environment and the application. The proposed toolkit insulates the application from context-sensing mechanisms through widgets. A context widget is a software component that provides applications with access to context information from their operating environment. The context widgets have a state (set of attributes that can be queried by applications) and a behavior (callbacks to the application when changes in the environment are detected). They are basic building blocks that manage sensing of the particular piece of context. A widget may perform one or more of the following roles: generator (acquire raw data from sensors), interpreter (abstract raw context into higher level information), and servers (collect, store, and interpret information from other widgets). The chapter describes three applications that use one or more of these context widgets:

- In/Out Board: This is the electronic equivalent of a simple in/out board that is found in offices. It is used to indicate which members of the office are currently in the building.
- Information Display: This displays information relevant to the user's location and identity, activated by the user's proximity. The information displayed changes to match the user, her research group, and location.
- DUMMBO (Dynamic Ubiquitous Mobile Meeting Board): This is an instrumented digitizing whiteboard that supports recording and replaying of informal and spontaneous meetings. Meeting recordings include whiteboard images as well as audio discussion. The chapter's authors present a revised version of DUMMBO in order to have recording triggered when two or more people are gathered around the whiteboard.

Schmidt and Strohbach [84]. The use of load sensing in everyday environments is considered as an approach to acquisition of contextual information in ubiquitous computing applications. While it is obvious that weight information is a useful context for the identification of objects, it is shown that load sensing can also be used to obtain positional information and interaction events on a given surface.

Weight is the most obvious contextual primitive that can be used for the identification of objects. The position of an object can as well be detected on a surface using load sensing. In this case, load cells are placed at the four corners of a table. Each load cell is connected to a commercial signal-conditioning unit, which in turn feeds a standard 16-bit Analog to Digital Converter (ADC) connected to a PC serial port. A program in Visual Basic periodically reads the measured load from the ADC and calculates the center of pressure based on a simple algorithm, thus obtaining the position of the object. An application that uses context information from load sensing called 'don't leave your things behind' reminds the user to take their objects with them when they leave the room. Whenever the placement of an object on the large table is recognized, the weight added to the table is stored together with the weight added on the floor (e.g., usually the person's weight). When the person leaves the floor (the overall weight is reduced by a certain amount) – the negative change of weight is used to check for an entry in the stored data set. If there is an entry – the person has put something down on the surface while entering – the software running on the PC provides an audio cue to remind the user to take his items with him. Another application tracks the position of a person in the space based on domain-specific knowledge about the location. Accumulating the tracking data over time offers a way of estimating the overall activity in the space.

2.3.3. *Context-Aware Application Life Cycle*

Brown [5] presents a new form of document, and the supporting software, which allows such applications to be created simply by building a new document. The motivation is to make the creation and use of these applications as easy as creating and using web pages. The stick-e document, a new form of document, which is presented in this chapter, is aimed at context-aware applications. It encompasses a metaphor whose purpose is to make such applications easier to create and under-stand. It covers a wide range of context-aware applications, but certainly not all, and removes the requirement for the creator of such applications needs to have computing skills; instead authorship just involves creating a stick-e document. Although different in purpose, stick-e documents share a number of similarities with WWW documents, and with hyper-documents in general. A stick-e document is built from smaller components, which are called stick-e notes. Each stick-e note consists of two parts: the content, as is normal for any document, and the context.

Dey and Salber [21] describe an architecture for supporting the software design and execution of context-aware applications. This architecture introduces the idea of context widgets for treating context as user input. An object-oriented approach has been used for the design of the architecture. The architecture consists of three main types of object: widgets, servers, and interpreters. A context widget provides an important abstraction to enable designers to use context without worrying about how the context was collected. It supports both the polling and notification mechanisms to allow components to retrieve current context information. A context server is used to collect the entire context about a particular entity, for example a person, in order to ease the job of an application programmer. A context interpreter is responsible for implementing the interpretation abstraction. By separating the interpretation abstraction from applications, reuse of interpreters by multiple applications is possible. An interpreter does not maintain any state information across individual interpretations, but when provided with state information, can interpret the information into another format or meaning. The use of the architecture is demonstrated through a complex application, the 'Conference Assistant' that assists a conference attendee, through a query interface, in a number of ways: from advising which presentations to attend based on his interests, to retrieving information about the conference once it is over.

Dey [22] focuses on a framework that makes it easier to design, build, and evolve context-aware applications. In this perspective the implemented 'context toolkit' is presented as well as a number of applications built using the context toolkit.

Finkelstein and Savigni [27] present a novel, reflection-based framework for requirements engineering for this class of context-aware applications. The chapter defines 'context awareness' as the ability of a particular service to adapt itself to a changing context. The framework addresses the difficulties in this field, such as changing context and changing requirements. The framework relies on the reflective approach. A reflective system maintains, at run time, data structures that materialize some aspects of the system itself. For the article's purposes, reflection means that an explicit, run-time representation of system behavior is maintained, which reifies the actual system behavior.

The proposed framework comprises the Goal, the Environment, the Context, the Requirement, the Service Description, and the Service components. The framework introduces each of these components, describes their functionality in the context-aware applications, and finally describes how each component affects the others. It also gives the following definitions:

- Goal is the objective the system should achieve through cooperation between agents in the software-to-be and in the environment.
- Environment is those real-world entities with which the machine interacts and the conditions under which the machine operates.
- Context is the reification of the environment.
- The Requirement represents one of the possible ways of achieving the Goal.
- Service Description is the meta-level representation of the actual real-world service.
- Service is the actual behavior as perceived by the user.

Klemke and Kanter [46,47] describes the SaiMotion project. In most context definitions, four dimensions of context are considered: the location of the user in either electronic (e.g., URL) or physical space, the identity of the user implying a user model with information about the user interests, preferences and knowledge, the time (day/night time working hours, weekend, etc.), and the environment (the task or activity in a current situation; other users). The process of information contextualization requires filtering, annotation, and aggregation of information contents. The identification of the relevant parameters that match the previously configured context values helps to perform this task. With the help of user, situation, and task monitoring it is possible to reduce the large amounts of information to a manageable level that matches the user's needs. However, the problem of privacy must be taken into account since users are not always willing to disclose their situation or parts of their situation (e.g., their location) to the SAiMotion-system or to other users.

Krause [50] identify and emphasize the necessity for quality of context (QoC) to enable and improve the automatic rating and processing of context information.

2.3.4. Context in GRID Computing

Many of the current GRID deployments have focused primarily on delivering high-performance computing, while future GRID applications are expected to enable complex application problem solving [35] for more diverse environments covering many more aspects of society such as health, genomics, new media, transport, energy, or public information systems. Due to the nature of the computational tasks GRID is expected to handle, non-trivial 'qualities of service' will have to be delivered. One approach to providing such functionalities is by augmenting applications running on hosts by an appropriate middleware. Alternatively, overlays can be used to provide augmented functionalities, mostly independent of properties of underlying infrastructures, in a less disruptive fashion than so-called integrated approaches. GRID computing research is now driven by the GRID Forum, which operates in a similar manner to the IETF. Open Grid Service Architecture for Distributed Systems Integration (OGSA) is under development. Open source OGSA middleware implementations (Globus Toolkit) are also continuously developed (see www.globus.org/ogsa).

2.3.5. Context-Aware Sensors' Computing

Harter *et al.* Hopper [34] describe a sensor-driven computing platform that collects environmental data and presents it in a form suitable for context-aware applications. The main components of the platform are the following:

1. A fine-grained location system, which uses ultrasonic techniques to locate and identify objects. Each object in the environment that is to be located has a small sensor tag attached to it that emits ultrasonic signal monitored by ultrasound

receivers placed at known points on the ceiling. Using the speed of sound in air, the position of the object to which the sensor tag is attached can then be deduced.
2. A rich data model that describes the essential real-world entities involved in mobile applications.
3. A persistent distributed object system, which presents the data model in a form accessible to applications. The software counterparts of real-world entities are implemented as persistent distributed objects using CORBA and an Oracle database.

2.3.6. Context-Aware Ontologies

The capabilities of different devices are best expressed using an ontology, against which the profiles of those devices are validated. W3C has a Device Independence activity, which works with CC/PP (Composite Capability/Preference Profiles) based on RDF [3]. A CC/PP profile contains CC/PP attribute names and associated values. The profile is structured to allow an entity to describe its capabilities by reference to a standard profile, accessible to a peer entity, and a smaller set of features that are in addition to or different than the standard profile. A CC/PP vocabulary consists of a set of CC/PP attribute names, permissible values, and associated meanings. CC/PP is compatible with IETF media feature sets (CONNEG) [39] in the sense that all media feature tags and values can be expressed in CC/PP.

DAML [36] is an ontology for expressing temporal aspects of the contents of web resources and time-related properties of web services. Modeling time is very important in context-aware architectures and applications, and therefore ontologies, such as the DAML-time ontology, are an essential component of such system.

In [38], the authors introduce another approach to use ontologies in the context of devices. An ontology-based description of functional design knowledge of engineering devices is presented. In the proposed model, generic concepts for representing the functionality of a device in the functional knowledge database are provided by the functional concept ontology, which makes the functional knowledge consistent and applicable to other domains.

A wireless world-related ontology is presented in [2]. The authors introduce an ontology to describe and discover services in an ad hoc networking environment such as Bluetooth. This ontology enables far better service discovery than simple UUID-based descriptions used in Bluetooth SDP system.

[40] is a sensor-based context ontology where each context is described using seven properties: (i) context type defines the category of the context; (ii) the context is the symbolic value of context type; (iii) the value property is the numerical value or feature describing context; (iv) an optional confidence property describes the uncertainty of context; (v) source property can be used to describe the semantic source of context; (vi) timestamp property defines the latest time when a context

occurred; and (vii) each context may have additional free attributes. RDF is used as the formal syntax for describing both structure and vocabulary of their ontology.

CORBA-ONT [16] is a collection of ontologies in the CoBrA architecture for smart spaces (e.g., intelligent meeting rooms, smart homes, and smart vehicles). Central to this architecture is an intelligent agent called context broker that maintains a shared model of context on behalf of a community of agents, services, and devices in the space and provides privacy protections for the users in the space by enforcing the policy rules that they define. Ontologies in CORBA-ONT are expressed in OWL. Key concepts in CORBA-ONT include ontologies about places (e.g., room, hallway), ontologies about agents (e.g., agent, person, role), ontologies about an agent's location context, and ontologies about an agent's activity context.

2.3.7. Context in Mobile Systems and Devices

In [44,60,79] the term context is considered to be application dependent or application independent. Context is further divided into network, user, and device context. Network and device context are divided into static, static in a cell, and dynamic context. The authors give examples regarding these divisions including context information that is about to be used. The challenge regarding context awareness is formulated as: to collect, process, distribute, and predict [77] the context data/information (e.g. to an application) in a scalable manner. In order to work with context information the following main points are stressed as require- ments: context management, context collection, context processing, and Dynamic Service Deployment. A further consideration is that cross-layer interfaces are important because of the fact that, without a cross-layer interface to layer two, it is not possible to collect information about the interface status [26,71–72]. The core building blocks of the architecture presented in [60] are: (i) Context Clients (CC) (e. g. an application running on an end-user device); (ii) Context Collection Points (CCP), running on both user devices and in the access network of providers; (iii) Context Service Adapters (CSA), acting as an interface between CCP and the CCs; and (iv) Service Deployment Framework (SDF), for controlling the provision of application-specific services in the network.

Karmouch *et al.* [43] defines a full context-aware architecture and system for ambient networks, which is a new class of networks that exploit the inherent heterogeneity seen in today's wireless networks in order to share diverse resources and services across different but cooperating networks. The ContextWare is an integrated infrastructure for sensing, processing, managing, and disseminating network context information to network entities and user-facing applications. It discusses how network-related context information should be utilized in ambient networks for the end user to fully experience the pervasiveness of a network and the research challenges arising from this utilization. The chapter also evaluates the benefits of employing context information and ContextWare concepts in ambient networks.

Gellersen [28–29] present the augmentation of mobile devices with awareness of their environment as context. Some definitions are given about context and context awareness: Context is what surrounds, and in mobile and ubiquitous computing the term is referred to as physical context. Situational context is what is inferred from the real world context, information acquired through sensors and processed to distill certain aspects of the surrounding world ('in a meeting,' 'driving in a car,' 'user is sleeping'). For example, if the sensors provide the information that the location is dark, room temperature, silent, indoors, that the time is 'night time,' and the user is horizontal with a specific motion pattern and the absolute position is stable, then the situation (context) is 'the user sleeps.' Devices may have direct or indirect awareness of context. In the case of indirect awareness, the entire sensing and processing occurs in the infrastructure while the mobile device obtains its context by means of communication. In contrast, a device has direct awareness if it is able to obtain context autonomously, (more or less) independently of any infrastructure. In this chapter, the devices have direct awareness of context. The chapter also defines a framework for the Active Artifacts:

- Autonomous awareness: Active Artifacts have sensors and perception methods embedded to assess their own state and situation independently of any infrastructure.
- Context sharing: Active Artifacts are augmented with the ability to communicate in order to make their context available within a local region of impact.
- Context use: any application in the local environment can use the context of Active Artifacts as a resource to enable enhanced functionality.

Khedr [49,51] present a technique for combining context information and agent technology to support spontaneous applications in an *ad hoc* network environment. The multi-agent system uses the Ad hoc Context-Aware Network (ACAN) infrastructure for context gathering, network setup, and application management. More specifically, the ACAN architecture consists of three layers: the mobility layer, the active ad hoc network layer, and the *ad hoc* applications layer. The mobility layer is network independent. Physical sensors in this layer gather information about the current environment and users in the system. The active *ad hoc* network layer is responsible for the auto-configuration and management of the network according to the context provided by the first layer. The *ad hoc* applications layer uses the context gathered from the first layer and the connectivity established by the second one to deploy applications spontaneously, discover services and users, and adapt its requirements according to the current situation. The multi-agent system is composed of four main components: sensor agents, context agents, discovery agents, and user agents. Each sensor of the

ACAN architecture acquires information about surrounding devices and services and delegates responsibility to a sensor agent that performs the tasks of data checking, data aggregation, communication with other sensor agents, and communication with the associated context agent. This last agent interprets the incoming information in order to extract specifications such as location, time, users, and values. Directory agents use the interpreted context to build a directory of available services and their attributes together with the service agent that represents each service. Service agents work as managers for the services they are in charge of. The paper also describes the Ontology Aware Service System (OASS) as the smart semantic model for supporting the representation of services, the interaction between the agents, and a common access to services that can adapt according to user profiles and service context.

2.3.8. Context Aware Communications

Knowledge Plane. DARPA started a project, called Knowledge Plane [13], which aims to add intelligence and self-learning to the network management. It aims to eliminate the unnecessary multi-level configuration. If one specifies the high-level design goals and constraints, the network should make the low-level decisions on its own. The system should reconfigure itself according to the changes in the high-level requirements. A distributed cognitive system, which permeates the network, is proposed that is called: knowledge plane (KP). Each networking element (end-node, router) has a KP. The KPs at a number of nodes interact with each other in order to keep themselves informed about global (network-wide) states and events. This interaction is also used to reconcile contradictory service levels and requirements. The KP must function in the presence of partial, inconsistent, and possibly misleading or malicious information. It must operate appropriately even if different stakeholders of the Internet define conflicting higher level goals. In order to meet these challenges, the authors suggest that cognitive techniques will be needed because analytical methods generally require precise and complete information. Nowadays, the network is usually divided into two architectural planes: a data plane and a control (or management) plane. The authors [13] believe that a new construct is needed instead of fitting knowledge into an existing plane. The KP would not move data directly, so it is not the data plane. However, unlike the control plane it tries to provide a unified view of the network rather than partition the world into managed segments. The KP integrates behavioral models and reasoning processes into a distributed networked environment. It supports the creation, storage, propagation, and discovery of information: observations (current conditions), assertions (high-level goals, constraints), and explanations (conclusions). Based on this information, the KP manages the actuators that change the behavior of the network components.

Kanter [47,48] introduces a novel, open, and scalable service architecture for context-aware personal communication. The proposed architecture deals with services for mobile users by supporting peer-to-peer service negotiation. Mobile agents [25] represent the users and other entities. Within the architecture, entities carry a context knowledge representation reflecting its capabilities, associated objects, and relationships between them. Entities can exchange context knowledge, merge it with existing knowledge, and interpret context knowledge in the end devices. Some middleware models are represented (JINI, UpnP, JXTA). These models allow devices to register services with a server, locate them, and use them. However, these protocols rely on external publication and knowledge of available objects and services. Kanter proposes the generic Extensible Service Protocol (XSP), allowing entities to extend their knowledge about context and available services in order to be able to use them. XSP is an extension of the SIP (Session Initiation Protocol). XSP allows us to keep *a priori* shared knowledge of service capabilities to a bare minimum and provide support to peers in order to discover, exchange, and reason about service knowledge. In this way, peers could establish connections to necessary services with a minimum of *a priori* knowledge. This chapter presents the HotTown prototype, which was implemented in order to demonstrate the feasibility of the architecture.

Plutarch [17] is an inter-networking architecture that makes heterogeneity, in the sense of network technology, explicit so that it may be exploited. A network composed of multiple network technologies is divided into contexts. Each context comprises a set of hosts, routers, switches, and network links. Within a context homogeneity of addresses, packet formats, transport protocols, and naming services is assumed. Communication between different contexts is enabled by so-called interstitial functions. An interstitial function provides a mapping between functionalities such as addressing, naming, routing, and transport between different contexts. The goal of Plutarch is to provide a set of compositional building blocks that allow the composition of heterogeneous networks to provide an end-to-end service. Within the Plutarch system, communication takes place between endpoints within contexts. Plutarch also defines (i) a context interface to be used by end systems, including an insert function, to add value to an instance of a context; (ii) an interstitial interface to allow end systems to interact with the special interstitial functions such as firewalls, NAT boxes, etc.; (iii) a Plutarch management service interface that gives access to a distributed service composed of multiple cooperating instances.

Schilit and Hilbert [80] focuses on a subset of context-aware computing named context-aware communication. Context-aware communication is defined as the class of applications that apply knowledge of peoples' context to reduce communication barriers. A two-dimensional space for such applications is suggested between 'context acquisition' and 'communication actions.' Along the 'acquisition' dimension, an application might ask people to manually enter their context, such as whether they are in a meeting or at lunch, or it may sense and infer a person's context with varying levels of autonomy and sophistication. Along the 'action' dimension, communication might be manually controlled. A set of context-aware communication applications is presented divided into five application types: routing, addressing, messaging, providing caller awareness, and screening.

2.3.9. Context-Aware Flows

Ocampo *et al.* [63–66] define flow context as any information that can be used to characterize the situation of a distinguishable stream of protocol data units, including information pertaining to the entities and circumstances that give rise to or accompany its generation at the source, affect its transmission through the network, and influence its use at its destination. Flows are the physical (or electronic) embodiment of the interaction between the user and networks, context information that characterizes these flows may be used to optimize or enhance this interaction. One approach being explored is to push flow context information, called flow context tags, along with the flow, under the assumption that the most interested consumers of such context (although not necessarily exclusively) would be the nodes along the flow's path. Flow context can be used to rapidly trigger services or adaptation within such overlays on short-lived flows, especially in the case of highly mobile hosts and in ad hoc networks; or to implicitly signal QoS requirements for media flows transported through media overlays.

References

1. Abstracts and Slides of the 'Workshop on Infrastructure for Smart Devices – How to Make Ubiquity an Actuality'. Web page. http://www.inf.ethz.ch/vs/events/HUK2kW/.
2. Avancha S, Joshi A, Finin T. 'Enhanced Service Discovery in Bluetooth'. *IEEE Computer* 2002; **35**(6):96–99.
3. Brickley D, Guha RV (ed.). 'RDF Vocabulary Description Language 1.0: RDF Schema'. W3C Working Draft, 2003. Work in progress.
4. Brown PJ, Bovey JD, Chen X. 'Context-aware applications: from the laboratory to the marketplace'. *IEEE Personal Communications* 1997; **4**(5): 58–64.
5. Brown PJ. 'The stick-e document: a framework for creating context-aware applications'. Electronic Publishing 1999; **8**: 259–272.
6. Brown PJ. 'Triggering information by context'. *Personal Technologies* 1998; **2**(1): 1–9.
7. Brown PJ. 'The electronic Post-it note: a model for mobile computing applications'. *Electronic Publishing* 1996; **9**(1): 1–14
8. Brown P, Burleson W, Lamming M, Rahlff O-W, Romano G, Scholtz J, Snowdon D. 'Context-awareness: some compelling applications'. *Proceedings the CH12000 Workshop on The What, Who, Where, When, Why and How of Context-Awareness*, April 2000.

9. Brumitt B, Meyers B, Krumm J, Kern A, Shafer S. EasyLiving: technologies for intelligent environments', Handheld and Ubiquitous Computing, September 2000.
10. Buchholz T, Kupper A, Schiffers S. 'Quality of Context Information: What is it is and why we need it'. In *proceedings of the 10th HP-OVUA Workshop*, Vol. 2003, Geneva, July 2003.
11. Sigrid B, PÔl M, Tore U. 'A simple Architecture for Delivering Context Information to Mobile Users. Position Paper at [IFSD00], 2000.
12. 'Context-Aware Applications Survey' www.hut.fi/~mkorkeaa/doc/context-aware.html.
13. Clark DD, Partridge C, Ramming JC, Wroclawski JT. 'A Knowledge Plane for the Internet' SIGCOMM2003, Karlsruhe, Germany, 2003.
14. Chen G, Kotz D. 'A survey of context-aware mobile computing research, Technical Report', TR2000-381, Department of Computer Science, Dartmouth College, November 2000.
15. Chen H, Finin T, Joshi A. 'An Intelligent Broker for Context-Aware Systems. In *Adjunct Proceedings of Ubicomp 2003*, Seattle, Washington, USA, October 12-15, 2003.
16. Chen H, Finin T, Joshi A. 'An Ontology for Context-Aware Pervasive Computing Environments'. In *IJCAI Workshop on Ontologies and Distributed Systems*, IJCAI'03, August, 2003.
17. Crowcroft J, Hand S, Mortier R, Roscoe T, Warfield A. 'Plutarch: An argument for network pluralism. *ACM SIGCOMM 2003 Workshops*, August 2003.
18. DeVaul RW, Pentland AS. 'The Ektara Architecture: The Right Framework for Context-Aware Wearable and Ubiquitous Computing Applications', The Media Laboratory, MIT, 2003.
19. Dey AK. 'Understanding and using context.' *Journal of Personal and Ubiquitous Computing* 2001; **5**(1): 4–7.
20. Dey AK, Abowd GD. 'Towards a better understanding of context and context awareness'. in *Workshop on the What, Who, Where, When and How of Context-Awareness*, affiliated with the 2000 ACM Conference on Human Factors in Computer Systems (CHI 2000), April 2000.
21. Dey AK, Salber D, Abowd GD, Futakawa M. 'An architecture to support context-aware applications'. GVU Technical Report: GIT-GVU-99-23, 1999.
22. Dey AK. 'Providing Architectural Support for Building Context-Aware Applications'. Thesis, Georgia Institute of Technology, 2000.
23. Dey A, *et al.* 'CyberDesk: A Framework for Providing Self-Integrating Ubiquitous Software Services'. Technical Report, GVU Center, Georgia Institute of Technology. GIT-GVU-97-10, May, 1997.
24. Dey AK. 'Context-Aware Computing: The CyberDesk Project'. AAAI 1998 Spring Symposium on Intelligent Environments, Technical Report SS-98-02 (1998), 51–54.
25. Foundation for Intelligent Physical Agents. FIPA Quality of Service Ontology Specification. Geneva, Switzerland. 2002. Specification number SC00094.
26. Fang Y, McDonald AB. 'Cross-layer performance effects of path coupling in wireless ad hoc networks: Power and throughput implications of IEEE 802.11 MAC'. In Proceedings 21st IEEE International Performance, Computing, and Communications Conference, April 2002, pp. 281–290.

27. Finkelstein A, Savigni A. 'A Framework for Requirements Engineering for Context-Aware Services'. In *Proceedings of STRAW 01 the First International Workshop From Software Requirements to Architectures*, 23rd International Conference on Software Engineering (2001).

28. Gellersen H-W, Schmidt A, Beigl M. 'Adding Some Smartness to Devices and Everyday Things'. In the *Proceedings of the Third IEEE Workshop on Mobile Computing Systems and Applications* (Monterey, CA, December 2000), ACM, 3–10.

29. Gellersen H-W, Schmidt A, Beigl M. 'Multi-Sensor Context-Awareness in Mobile Devices and Smart Artefacts'. *Proceedings of UBICOMP 2001*, Atlanta, GA, USA, September 2001.

30. Catalin G, Julien C, Huang Q. 'Network Abstractions for Context-Aware Mobile Computing'. http://citeseer.nj.nec.com/roman01network.html

31. Gray PD, Salber D. Modelling and Using Sensed Context Information in the Design of Interactive Applications. In Proceedings 8th IFIP Working Conference on Engineering for Human-Computer Interaction (EHCI'01), May 2001.

32. Hohl F, Kubach U, Leonhardi A, Rothermel K, Schwehm M. 'Next Century Challenges: Nexus—An Open Global Infrastructure for Spatial—Aware Applications'. *Proceedings of the Fifth Annual ACM/IEEE International Conference on Mobile Computing and Networking (MobiCom'99)*, Seattle, Washington, USA, August 15–20, 1999 Imielinski, T Steenstrup, M (Eds.), ACM Press, 1999, pp. 249–255.

33. Hong JI, Landay JA. 'An Infrastructure Approach to Context-Aware Computing'. In Human-Computer Interaction, 2001, Vol. 16, 2001.

34. Harter Hopper A, Steggles P, Ward A, Webster P. 'The anatomy of a context-aware application'. In *Proceedings of MOBICOM 1999*, pp. 59–68.

35. Hightower J, Borriello G. 'Location systems for ubiquitous computing'. *IEEE Computer* 2001; 57–66.

36. Helin H.'Supporting Nomadic Agent-based Applications in the FIPA Agent Architecture'. PhD. Thesis, University of Helsinki, Department of Computer Science, Series of Publications A, No. A-2003-2. Helsinki, Finland, January 2003. H. Helin and M. Laukkanen. Wireless Network Ontology. In *Proceedings of the Wireless World Research Forum 9th Meeting*. ZÏrich, Switzerland, July 2003.

37. Jean K, Galis A, Tan A. 'Context-Aware GRID Services: Issues and Approaches'. IEEE International Conference on Computational Science (ICCS) 2004, APGAC'04. 'First International Workshop on Active and Programmable Grids Architectures and Components', KrakÆw, Poland, June 7–9, 2004; www.ics.agh.edu.pl/apgac2004/; - www.cyfronet.krakow.pl/iccs2004/index.html

38. Kitamura Y, Kasai T, Mizoguchi R. Ontology-based Description of Functional Design Knowledge and its Use in a Functional Way Server, Proceedings of the Pacific Conference on Intelligent Systems 2001.

39. Klyne GA. 'Syntax for Describing Media Feature Sets'. RFC 2533, 1999.

40. KorpipÏÌ P, MÌntyjÌrvi J. 'An Ontology for Mobile Device Sensor-Based Context Awareness'. In Fourth International and Interdisciplinary Conference on Modeling and Using Context (CONTEXT 2003): 451–458. Stanford, California (USA), June 23–25, 2003.

41. Korpipìì P, Mìntyjìrvi J, Kela J, Kerìnen H, Malm, E-J. 'Managing Context Information in Mobile Devices'. *IEEE Pervasive Computing* 2003; **2**(3):42–51.

42. Kleinrock L. Nomadicity: 'Anytime, anywhere in a disconnected world'. *Mobile Networks and Applications* 1996; **1**: 351–357.

43. Karmouch A, Galis A, Giaffreda R, Kanter T, Jonsson A, Karlsson AM, Glitho R, Smirnov M, Kleis M, Reichert C, Tan A, Khedr M, Samaan N, Heimo L, Barachi ME, Dang J. 'Contextware Research Challenges in Ambient Networks' – ISBN 3-540-23423-3, Springer- Verlag Lecture Notes in Computer Science- IEEE MATA 2004-20-22 October 2004, Florianopolis, Brazil - www.ic.unicamp.br/mata04/

44. Kanter T, Lindtorp P, Olrog C, Maguire GQ. 'Smart delivery of multimedia content for wireless applications'. *Mobile and Wireless Communication Networks* 2000; 70–81.

45. Klemke R, Nick A. 'Case Studies in Developing Contextualising Information Systems'. in: CONTEXT-01—Third International and Interdisciplinary Conference on Modeling and Using Context, Dundee (Scotland), July 27–30, 2001.

46. Klemke R. 'Context Framework—An Open Approach to Enhance Organisational Memory Systems with Context Modelling Techniques'. In PAKM-00: Practical Aspects of Knowledge Management, Proceedings 3rd International Conference, Basel. Switzerland, 2000.

47. Kanter TG. 'Hottown, enabling context-Aware and extensible mobile interactive spaces'. Special Issue of IEEE Wireless Communications and IEEE Pervasive on 'Context-Aware Pervasive Computing'. October 2002, pp. 18–27.

48. Kanter TG, Gerald QM, Jr., Smith MT. Rethinking Wireless Internet with Smart Media-http://psi.verkstad.net/Papers/conferences/ nrs01/nrs01-theo.PDF.

49. Khedr M, Karmouch A. 'Exploiting SIP and agents for smart context level agreements'. 2003 IEEE Pacific Rim Conference on Communications, Computers and Signal Processing, Victoria, BC, Canada, August 2003.

50. Krause M, Hochstatter I. 'Challenges in Modeling and Using Quality of Context' – ISBN10 3-540-29410-4, Springer- Verlag Lecture Notes in Computer Science- IEEE MATA 2005 – 17-19 October 2005, Montreal, Canada.

51. Khedr M, Karmouch A, Liscano R, Gray T. 'Agent-based context aware ad hoc communication'. In Proceedings of the 4th International Workshop on Mobile Agents for Telecommunication Applications (MATA 2002), Barcelona, Spain, October 23-24, 2002, pp. 292–301.

52. Kempf J. 'Problem Description: Reasons For Performing Context Transfers Between Nodes in an IP Access Network-RFC 3374; 2002.

53. Korkea-aho, M. 'Context-Aware Applications Survey'. Internetworking Seminar (Tik-110.551), Spring 2000, Helsinki University of Technology.

54. Hegering HG, Kupper A, Popien CL, Reiser H. 'Context-Aware Services in Ubiquitous Environments'. In IFIP/IEE Workshop on Distributed Systems: Operations and Management, DSOM 2003, Heidelberg, October 2003, ISBN 3-540-20314-1 Lecture Notes in Computer Science, Springer-Verlag.

55. Leonhardi A, Kubach U. 'An Architecture for a Universal, Distributed Location Service'. *Proceedings of the European Wireless '99 Conference*, Munich, Germany. ITG Fachbericht, VDE Verlag, 1999, pp. 351–355.

56. Long S, Kooper R, Abowd GD, Atkenson CG. 'Rapid Prototyping of Mobile Context-Aware Applications: The Cyberguide Case Study'. *Proceedings of the Second Annual International Conference on Mobile Computing and Networking*, November 1996, Rye, New York, United States. pp. 97–107.

57. Mandato D, Kovacs E, Hohl F, Amir-Alikhani H. 'CAMP: 'A context-aware mobile portal'. *IEEE Communications Magazine* 2002; **40**(1): 90–97.

58. Nakamura T, Matsuo N, Itao T. 'Context Handling Architecture for Adaptive Networking Services'. *Proceedings of the IST Mobile Summit* 2000.

59. Eisenhauer M, Klemke R. 'Contextualisation in Nomadic Computing'. Ercim News, Special Issue in Ambient Intelligence, October 2001.

60. Mendes P, Prehofer C, Wei Q. 'Context management with programmable mobile networks'. *IEEE Computer Communication Workshop*, 2003.

61. Oppermann R, Specht M. 'A Context-sensitive Nomadic Information System as an Exhibition Guide'. Handheld and Ubiquitous Computing Second International Symposium.

62. Orr RJ, Abowd GD. 'The Smart Floor: A Mechanism for Natural User Identification and Tracking'. *Proceedings of the 2000 Conference on Human Factors in Computing Systems*, The Hague, Netherlands, April 1–6, 2000.

63. Ocampo R, Galis A, Todd C. 'Triggering Network Services Through Context-Tagged Flows'. *Proceedings of the 5th International Conference on Computational Science (ICCS'05)*, Atlanta, Georgia, USA, May 2005.

64. Ocampo R, Galis A, De Meer H, Todd C. 'Flow Context Tags: Concepts and Applications'. IFIP TC6 Conference on Network Control and Engineering for QoS, Security and Mobility (NetCon'05), Lannion, France, November 2005.

65. Ocampo R, Galis A, De Meer H, Todd C. 'Supporting Mobility Adaptation Through Flow Context'. *2nd International Workshop on Mobility Aware Technologies and Applications (MATA 2005)*, Montreal, Canada, October 2005.

66. Ocampo R, Galis A, De Meer H, Todd C. 'Implicit Flow QoS Signaling Using Semantic-Rich Context Tags'. *Proceedings of the 13th International Workshop on Quality of Service (IWQoS 2005)*, Passau, Germany, June 2005.

67. Ocampo R, Cheng L, Galis A. 'ContextWare Support for Network and Service Composition and Self-Adaptation'; IEEE MATA 2005. Mobility Aware Technologies and Applications, - Service Delivery Platforms for Next Generation Networks; Springer ISBN-2 553-01401-5; 17–19 October 2005, Montreal, Canada; www.congresbcu.com/mata2005/.

68. Priyantha N, Chakraborty A, Balakrishnan H. 'The Cricket Location-Support System'. In *Proceedings of the Sixth Annual International Conference on Mobile Computing and Networking* (Boston, MA, August 2000), ACM, 32–43.

69. Pascoe J, Ryan N, Morse D. 'Issues in developing context-aware computing'. In Proceedings First International Symposium on Handheld and Ubiquitous Computing (HUC'99), 1999.

70. Pascoe J. The Stick-e Note Architecture: Extending the interface beyond the user. In Proceedings 2nd International Conference on Intelligent User Interfaces, January 1997.

71. Raman B, Bhagwat P, Seshan S. 'Arguments for cross-layer optimizations in bluetooth scatternets'. In Proceedings of Symposium on Applications and the Internet (SAINT), January 2001.

72. Raman B, Katz P, Joseph A. 'Universal Inbox: Providing Extensible Personal Mobility and Service Mobility in an Integrated Communication Network'. In *Proceedings of the IEEE Workshop on Mobile Computing Systems and Applications* (Monterey, CA, December 2000), IEEE, 95–106.

73. Starner T, Kirsh D, Assefa S. 'The locust swarm: An environmentally-powered, networkless location and messaging system'. In Digest of Papers. 1st International Symposium on Wearable Computers, (1997), 169–170.

74. Starner T, Kirsh D, Pentland A. 'Visual context awareness in wearable computing'. In Digest of Papers. 2nd International Symposium on Wearable Computers (1998), 50–57.

75. Schilit B, Adams N, Want R. 'Context-Aware Computing Applications'. *Proceedings of the 1st International Workshop on Mobile Computing Systems and Applications*, 1994, pp. 85–90.

76. Schmidt A, et al. 'Advanced interaction in context'. First International Symposium on Handheld and Ubiquitous Computing (HUC99), Karlsruhe, Germany, September 1999, LNCS 1707, Springer-Verlag.

77. Samann N, Karmouch A. 'An evidence-based mobility prediction agent architecture'. in proceedings of the 5th International Workshop on Mobile Agents for Telecommunication Applications (MATA2003), Marrakesch, October 2003, ISBN 3-540-20298-6 - Lecture Notes in Computer Science, Springer-Verlag.

78. Salber D, Dey AK, Abowd GD. 'The context toolkit: Aiding the development of context-enabled applications'. In *Proceedings of CHI'99, May 1999*.

79. Schilit WN. 'A system architecture for context-aware mobile computing'. Ph.D. thesis, Columbia University, 1995.

80. Schilit BN, Hilbert DM, Trevor J. 'Context-aware communication'. *IEEE Wireless Communications* 2002; **9**(5): 46–54.

81. Schilit BN, Theimer MM. 'Disseminating active map information to mobile hosts'. *IEEE Network* 1994; **8**(5): 22–32.

82. Schmidt A, Beigl M, Gellersen HW. 'There is more to context than location'. *Computers & Graphics Journal, Elsevier* 1999; **23**(6): 893–902.

83. Schmidt A, van Laerhoven K. 'How to build smart appliances?'. *IEEE Personal Communications* 2001; **8**(4): 66–71.

84. Schmidt A, Strohbach M, Van Laerhoven K, Friday A, Gellersen H-W. 'Context Acquisition based on Load Sensing'. In *Proceedings of Ubicomp 2002* Boriello G, Holmquist LE (Eds). Lecture Notes in Computer Science, Vol. 2498, ISBN 3-540-44267-7; Gîteborg, Sweden. Springer-Verlag, September 2002, pp. 333–351.

85. Tuulari, Esa. 'Context aware hand-held devices' Espoo, VTT Electronics, 2000. VTT Publications.

86. Want R, Hopper A, Falcao V, Gibbons J. 'The active badge location system'. *ACM Transactions on Information Systems* 1992; **10**(1): 91–102.

87. Want R, Schilit BN, Adams NI, Gold R, Petersen K, Goldberg D, Ellis JR, Weiser M. 'The ParcTab ubiquitous computing experiment'. Xerox PARC Computer Science Laboratory Tech Report CSL-95-1, March 1995.

88. Want R, Schilit BN, Adams NI, Gold R, Petersen K, Goldberg D, Ellis JR, Weiser M. 'An overview of the ParcTab ubiquitous computing experiment'. *IEEE Personal Communications* 1995; **2**(6): 28–43.

89. Want R, Schilit BN. 'Expanding the horizons of location-aware computing'. Guest Editor's Introduction. *Computer* 2001; **34**(8): 31–34.

90. Winograd T. 'Architecture for Context'. *Human Computer Interaction Journal*, 2001, Vol. 16, pp. 401–419.

91. Yang K, Galis A. 'Network-centric context-aware service over integrated WLAN and GPRS networks' 14th IEE International Symposium On Personal, Indoor And Mobile Radio Communications, September 2003.

92. Yang K, Galis A. 'Policy-driven mobile agents for context-aware service in next generation networks'. *IFIP 5th International Conference on Mobile Agents for Telecommunications*, Marrakesch, October 2003, ISBN 3-540-20298-6 - Lecture Notes in Computer Science, Springer-Verlag.

93. Yau SS, Karim F. 'An Adaptive Middleware for Context-Sensitive Communications for Real-Time Applications in Ubiquitous Computing Environments'. Real-Time Systems, The International Journal of Time-Critical Computing Systems, Kluwer Academic Publishers, Dordrecht, The Netherlands, Vol. 26, no. 1, pp. 29–61, Jan 2004.

94. Yau SS, Karim F. 'Context-Sensitive Object Request Broker for Ubiquitous Computing Environments'. *Proc. 8th IEEE Workshop on Future Trends of Distributed Computing Systems (FTDCS 2001)*, Bologna, Italy, Oct 31-Nov 2, 2001.

95. Yau SS, Karim F. 'A Lightweight Middleware Protocol for Ad Hoc Distributed Object Computing in Ubiquitous Computing Environments'. *Proc. 6th IEEE Intl. Symposium on Object-Oriented Real-Time Distributed Computing (ISORC 2003)*, pp. 172–179, May 2003.

3

The Service Life Cycle Functional Architecture

This chapter reviews the main research results relevant to the service life cycle of Context-Aware Services. The major phases of the life cycle, the 'epochs,' derived from the order of execution of different aspects of the service life cycle, are introduced. Each epoch is described and its functional components are identified. The functional components, which constitute the functional architecture of the system that must ensure the whole service life cycle, are then described in detail.

3.1. Introduction

According to ETSI [1,2], the service life cycle is the description of phases and activities involved during the complete life of any service, in a service-independent manner. It is considered the basis for defining the possible behavior of a service at all times, the phases identified covering all aspects of a service life, including its 'death.' The service life cycle is composed of several phases, with specific activities that can be carried out by one or more actors in each phase. Although the above concepts were grown around the Intelligent Networks (IN), it is intended that this framework applies to all possible services. Another view is given by W3C in respect to the service life cycle of Web Services [3]. In this case, the service is characterized at a given point of its life by a state. Changes between states are characterized by specific transitions. In addition, the states can be subdivided into substates where, in turn, there are the appropriate transitions. The service life cycle as a whole is the scope of the discipline called Service Engineering [4–9].

The service life cycle includes the procedures and steps intended for the design, implementation, activation, operation, and withdrawal of a service [10–12]. This set of procedures, which can be understood as arranged in sequence, are often divided

Fast and Efficient Context-Aware Services   Danny Raz, Arto Tapani Juhola,
Joan Serrat-Fernandez, Alex Galis   © 2006 John Wiley & Sons, Ltd

into two functional areas, namely the service creation and the service management. The boundary between these two areas is established when the service is ready to be deployed. Up to that moment, the processes taking place belong to the category of service creation whilst from that moment they belong to the category of service management. All the above steps may be decomposed in other substeps. In particular, the operational phase includes activities like deactivation, reactivation, removal, migration, replication, and updates. Some of these activities may be supported by others that fall outside the actual one. For instance, service migration may entail the deactivation followed by installation of code in other hosts or execution environments, although service installation is not normally considered to be an operational step.

Once created, services may be installed in different distribution points ready to be downloaded into the execution environments as soon as they are required. The criteria to select these distribution points are diverse, and therefore this activity must be under the scope of the service management system. This is a fundamental difference from previous notions of software installation linked to the software installation in specific hosts. Management applications must be able to install and remove services [13–15]. Furthermore, a list of currently installed services must be accessible. Required information includes data identifying the code, such as the service's name and its vendor, version information, and the service interfaces provided by the code.

The first step in the service operational phase is its activation. Activating a service means that the service's code is executed in an execution environment. Here again the management system should decide in which execution environment the code will be service activated. It is worth mentioning that the service itself may decide on when to carry out a deactivation, suspension, and resumption. Most services need a certain set of configuration data. A service must be able to store and (on restart) retrieve its settings. This is not a trivial issue as services can be started on arbitrary and varying execution hosts. As hosts may have different setups, the traditional solution of a configuration file in a well-known location is not viable. The infrastructure, therefore, has to provide some other means of handling configuration data in a location-independent way.

3.2. Service Life Cycle Model for Context-Aware Services

Context-Aware Services (CAS) are not much distinct from other types of services as far as the service life cycle is concerned. Therefore, we adopt the same basic concepts exposed in the previous section, adapt the terminology, and focus on the phase and activities that have a major impact and highlight the most particular aspects in our context. Consequently, the entire life cycle of a CAS is divided into the four distinct epochs depicted in Figure 3.1.

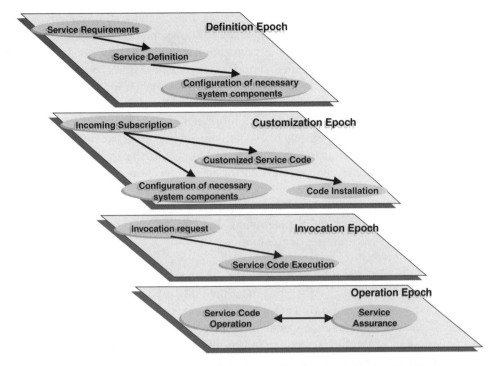

Figure 3.1   Proposed Context-Aware Services Life Cycle Model.

The service is given birth during the definition epoch, before which it has only existed as an idea based on a set of requirements derived from market analysis and business considerations. The service idea is translated into a technical description of the new service, called the Service Definition, encompassing all the functionality necessary to fulfill the requirements together with a specification of the system configurations deemed to be necessary to provision the service. The definition epoch finishes with the dispatch of the service definition and the configuration specifications for the new service to the appropriate components of the provisioning system.

The customization epoch of the new service takes place at the granularity of incoming subscription requests from potential consumers of this service. Each new subscription request initiates a consumer-specific customization epoch leading to the establishment of an agreement between the consumer and the service provider. This agreement details the terms of service use, including any service customizations. Based on the agreement, a service subscription is established and the customized service code is produced and subsequently installed in the infrastructure that will support its activation and operation. In addition, any new infrastructure configurations required to support the newly established subscription are effected. In the case of services offered to a single consumer (e.g., the network itself), or to a group of consumers not needing an explicit subscription, the customization epoch is not activated and the invocation epoch immediately follows the definition epoch.

After successful completion of the customization epoch, the subscribed service is ready for use and may be instantiated and triggered in two ways. The first way is in response to an incoming invocation request, originating either from the user (using a signaling protocol) or from the infrastructure (specific alarms). An invocation request triggers a series of actions leading ultimately to execution of the code of one or more services at proper network locations. Service code execution will take into account all relevant information encapsulated within the invocation request. Alternatively services may be designed such that no invocation request is required for their use. In such cases the service code is either positioned and executed immediately after the customization epoch (or the definition epoch in the case of customization epoch absence) or positioning and execution is based on a service schedule detailed within the service definition.

A successful invocation epoch is followed by the operation epoch, during which the service delivers the required functionality. Operation of the service is supported by infrastructure mechanisms that implement the APIs used by the service (e.g., for retrieving context information). Throughout the operation epoch, efficient service delivery is ensured by assurance mechanisms that monitor performance and change the configuration of the infrastructure, and of the service itself, as required.

An important part of the operation epoch is the cessation of the execution of code at certain locations at certain points in time and restarting of code at other locations. This is necessary to support mobility without flooding the whole network with pieces of personalized code, a strategy that aside from the sheer volume of the associated traffic would amount to a SW-maintenance nightmare.

The service layer functional architecture must support all epochs of the service life cycle and all aspects of the functionality these epochs encompass. In addition, it must be sufficiently flexible and generic to allow automated definition and provisioning of context-aware services, taking into account the complexity and variability of context information, and sufficiently dynamic to manage these services efficiently. Figure 3.2 presents the major functional areas of the service layer that are required to fulfill the above-mentioned requirements.

The Service Creation functional block is responsible for handling service definition and customization, and is therefore active during the definition and customization epochs. It delivers to the Policy-Based Service Management functional block the customized service code and the policies pertinent to the service's provisioning.

The Policy-Based Service Management functional block is responsible for configuration and performance management of the created services. The policies that influence the management operations are set during the Service Definition epoch. Configuration management is active during the service operation epoch and decides the terms of service execution. Performance management is active also during the service operation epoch and ensures correct operation of executing services.

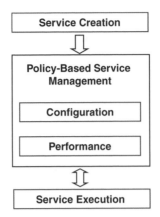

Figure 3.2 Service Life Cycle Decomposition into Major Functional Areas.

The Service Execution functional block is responsible for executing and support-ing service operation, and therefore is active during the service operation epoch. It executes services after receiving invocation requests, with specific terms, from the Service Management block. It supports service operation by providing several APIs, which make the infrastructure capabilities available to the services.

Finally, the Service Execution communicates monitoring data to the Performance Management block. By further analyzing the major functional blocks presented above we have defined with the complete Service Layer high-level functional architecture. This architecture is presented by Figure 3.3. Service creation function-ality is realized by the CAS Authoring, the Service Customization, and the Code and Policies Generation Engine functional components.

The CAS Authoring component, active during the definition epoch of the service life cycle, is responsible for producing a coherent and complete 'technology independent' service definition. The logic of the component is human driven, assuming a CAS administrator who creates the service definition using the CAS Authoring tools. These tools elevate the abstraction level of the service definition process, thus rendering it human friendly, and guide the definition process, preventing errors and inconsistencies.

The result of the CAS Authoring component is fed to the Code and Policies Generation Engine that translates the service definition into service code and configuration policies. This process takes as input the service definition and produces (a) the technology-specific service code, fitting the execution environment of the service and (b) sets of policies for the components of the management system. These results are disseminated to the appropriate components, in order to obtain the necessary configurations pertinent to the service to be performed. The service definition epoch completes with the successful production of these

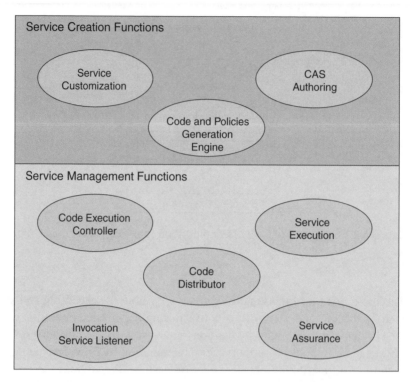

Figure 3.3   Service Layer Functional Architecture.

configurations. The Service Customization component, active during the customization epoch, is responsible for establishing subscription agreements with new consumers of existing subscription-based services. Following successful subscription establishment the derived customizations are forwarded to the Code and Policies Generation Engine.

The Code and Policies Generation Engine is active during both the customization epoch and the definition epoch. During the customization epoch it produces and disseminates the customized service code, that is, the 'standard' service code plus the customizations derived from the consumer's subscription. Additionally, it produces and disseminates the policies required to realize the system component configurations necessary to deploy this subscription.

Policy-Based Service Management functionality is performed by the Code Distributor, the Code Execution Controller, the Invocation Service Listener, and the Service Assurance components.

After its creation, the customized service code is installed at specific points of the infrastructure by the Code Distributor component. Selection of these points takes into account the service management policies. Following code installation, Code Distributor is responsible for informing the Code Execution Controller of the URLs of the installed code. Maintenance of the installed code is also a part of this component's functionality.

It is to be noted that initially the code is not necessarily installed into the nodes they will actually execute, but in one or more intermediate storage points. With many services the target nodes for the code can be determined during the service operation epoch only, being dependent, for example, on the location of (mobile) hosts.

To solve the problem of assigning code to the correct locations just-in-time, a generic mechanism was devised, to be described later in this section.

The Invocation Service Listener component, activated during the service invocation epoch, is responsible for receiving triggers in the form of various protocol signals or infrastructure alarms. It is a distributed component that spans the whole network, positioning listeners that are specialized for a specific type of trigger at the points where this trigger can be captured. After capturing a trigger, the listeners encapsulate the trigger information into a message of specific form and forward it to the Code Execution Controller. When more than one Code Execution Controllers is employed by the system, the selection amongst them is made based on management policies.

The Code Execution Controller receives trigger messages from the Invocation Service Listener components. Every message is decomposed into a set of parameters representing the trigger information. The Code Execution Controller is configured with a set of policies, which are set during the creation of the service and the establishment of each service's subscription. These policies, when evaluated with the parameters deduced from the trigger, will result in activation requests for specific customized services (one or more) and on specific terms applicable to these activations. The Service Execution is then passed the resulting customized service activation requests, along with their terms, and undertakes execution of these services. Because the execution environment is distributed, the exact node(s) where the services will be executed need to be selected, again based on predefined management policies, in order to forward the activation requests to the correct nodes. For scalability, management, and resilience reasons, more than one Code Execution Controller may exist. Each one may serve only a subset of the Invocation Service Listeners or a subset of services.

The execution environment active during the operation epoch, provides a platform for service code execution and fulfillment of the operational needs of this code. The execution environment implements APIs that offer services access to the capabilities of the infrastructure. During the customization epoch, all new services are designed to access these capabilities using the APIs provided. In addition, the execution environment communicates with the Service Assurance component to dynamically exchange monitoring and configuration data in order to maintain correct service operation.

The Service Assurance component, active during service operation epoch, is responsible for monitoring the performance of operating services and, in the case of performance degradation, taking appropriate corrective measures. Monitoring is done at the granularity of the system, of individual service code, and of the infrastructure mechanisms supporting the operation of the services. The results of monitoring are fed to the Performance Asserter part of the Service Assurance component that, based on the management policies that influence its behavior, will decide the enforcement of proactive and reactive measures. Proactive measures aim to achieve the best system configuration to meet future requirements. One example of proactive measures is the production of policies that alter the behavior of the Code Execution Controller to direct future service activation requests to less utilized nodes of the execution environment, or alternatively to deny specific service activations. Reactive measures aim to rectify current system operation in order to improve the performance of currently executing service code. One example of a reactive measure is the real-time termination or reconfiguration of individual operating service code in order to free system resources for other services.

Figure 3.4 presents the reference points for information exchange between the functional components of the Service Layer as presented earlier in Figure 3.3.

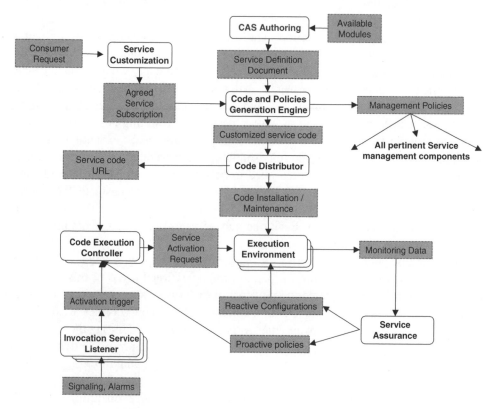

Figure 3.4   Service Layer Reference Points.

3.3. Service Creation

Of all the stages of the service life cycle, service creation is one of the most abstract and general, since there are not many detailed guidelines (advisory statements) available on how to structure each of its phases. Furthermore, it is also one of the most important since it determines the efficiency with which the services will be developed and thus the success of service providers in a highly competitive market [22]. As pointed out by the same authors, the service creation process does not need to follow a waterfall model in which each activity is done once for the entire set of service requirements, because the traditional waterfall model for software development with its sequential phases is inadequate to support the development of telecommunications services.

Service creation is widely studied in the telecommunication community (e.g., ITU-T, ETSI, TINA-C, RACE, ACTS, or EURESCOM). All these studies adopt a common approach that is the definition of an architecture and a methodological framework with its support. Thus, the IN and TINA architectures have been defined [2,23], and the literature describes plenty of different implementations [24–27]. The methodological aspect is described by the service creation environment (SCE) concept. This enables one to unify the process of service creation by defining a role model, a service life cycle model, and a set of methods and tools that support the activities of all the roles.

In line with the efforts of the above-mentioned initiatives, we can mention different solutions to the problem of service creation. A methodology to simplify the process of service creation is proposed in Reference [28]. A set of broadband service-independent building blocks are designed and used to create and customize broadband services. During service execution time, a service agent interprets a building block graph and executes its procedures, which are all downloaded from the service provider. The work presented in Reference [29] is a specialization of the service life cycle to web services particularly putting the emphasis on the service creation part, showing the need to use design patterns and Web requirements analysis techniques and methodologies. A structured generic approach to the service creation problem and solution is presented in Reference [22] whilst the Cadenus approach [30] provides service creation and configuration in a dynamic way through its own framework. Other relevant work related to issues in service creation can be found in references [4,17,21,28,31].

3.3.1. CAS Authoring

The objective of this function is twofold: first of all, to produce a coherent and complete service definition that will be the basis for creation and provisioning of the service; secondly, to provide a tool to assist the CAS creation administrator to compile the service definition.

Concerning the above-mentioned specification tool, it is worth stating that it should elevate the abstraction level of the service definition process, thus rendering it human friendly and technology independent. In addition, it must present all the available infrastructure capabilities from the service creation perspective. Finally, it must guide the definition process so as to prevent potential errors and inconsistencies in the service definition.

As shown in Figure 3.4, the CAS Authoring component accepts as input the available modules representing the system capabilities for constructing CAS. These modules must be based on a preestablished CAS modeling approach. The output of this component is the *Service Definition Document*, containing an implementation technology-independent service definition describing all aspects of a service's creation, provisioning, operation, and management.

The detailed functionality of the CAS Authoring component is put in place when a CAS administrator logs in, and ends with the production of a service definition document as presented by Figure 3.5.

Login to the CAS authoring tools will be permitted after successful authentication and authorization of the administrator's credentials, and completion of a list of permitted actions for this administrator. The list of actions consists of options such

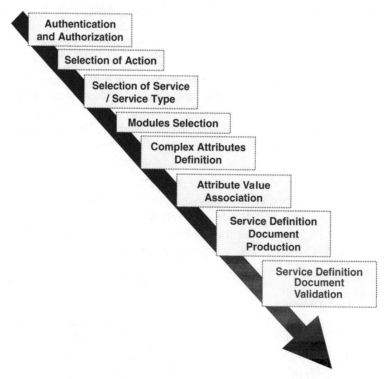

Figure 3.5 CAS Authoring Component Functionality.

as CAS Creation, CAS Deletion, and CAS Modification. Selecting an option from the above list leads to the compilation of a new list of options applicable to this selection. For example, if CAS Creation is selected, a list of potential service types is presented. After all initial options have been selected, the modules to be included in the service definition must be selected. If any of the attributes of the selected modules are complex, they must be defined according to their corresponding complex attribute schema. The attribute-value association process then must follow. The input attributes of the selected modules must be associated with values of matching types. The values can be fixed either by the administrator or by other selected modules. Caution must be taken to avoid loops. This can be achieved by only allowing a module's output attribute to be associated with a value after all the module's input attributes have been associated with values. After completion of the attribute-value association process, during which the results of all the previously employed processes are taken into account, the Service Definition Document is produced as an XML document, which is then validated against the rules pertinent to its service type. If inconsistencies are detected then the administrator is informed and prompted to modify the service definition. The end of the definition process is reached when a valid Service Definition Document has been produced and forwarded to the Code and Policies Generation Engine component.

The functionality of the CAS creation component can be realized by the components presented in Figure 3.6.

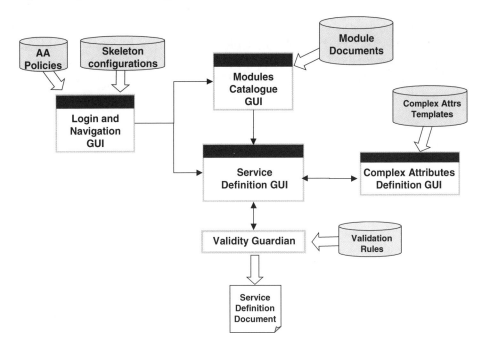

Figure 3.6   CAS Authoring Component Architecture.

The Login and Navigation GUI is responsible for handling the administrator's login, and the presentation and selection of the available CAS authoring options. It is dynamically configured by authentication and authorization policies and by skeletons that define the available CAS authoring options and their inter-relationships. Based on the options selected by the administrator, a list of modules that make up the service is deduced from the pool of available modules. The Modules Catalog GUI presents these modules to the administrator and facilitates their discovery and selection by providing a comprehensive search engine. The selected modules are transferred as graphical objects to the Service Definition GUI. If the selected modules exhibit complex attributes, these can be defined using the Complex Attribute Definition GUI. This GUI is configured with a schema that dictates the valid structure for each complex attribute and is used to guide the administrator through the attribute's definition. The attribute-value association process is performed utilizing the graphical environment provided by the Service Definition GUI. At the end of this process, the resulting complete graphical representation of the service definition is translated into the XML code that constitutes the Service Definition Document. This document is validated by the Validity Guardian, using the validation rules for the type of service selected, and any detected inconsistency is reported back to the administrator through the Service Definition GUI. The final valid Service Definition Document is forwarded to the Code and Policies Generation Engine component.

The CAS Authoring component can be implemented as a stand-alone application, integrating all necessary tools that facilitate the creation of services and offer the administrator an easy-to-use graphical interactive interface. The modeling of CAS is based on XML and on the XML schema specifications. The GUIs can be implemented using Java technology, including swing libraries 2nd XML parsers. The repositories containing the required configuration information (policies, module documents, etc) can be implemented using database technology. The interface with the Code and Policies Generation Engine can be implemented using SOAP.

3.3.2. Service Customization

The objectives of the Service Customization component are to establish subscription agreements with service consumers for existing services and to forward the customizations derived for the established subscription to the Code and Policies Generation Engine.

The Service Customization component accepts as input the consumer's service request, containing all the service customizations necessary to meet the consumer's individual needs. After appropriate processing it delivers a document containing the terms of service use agreed between the provider and the consumer, including the consumer's service customizations (see Figure 3.4).

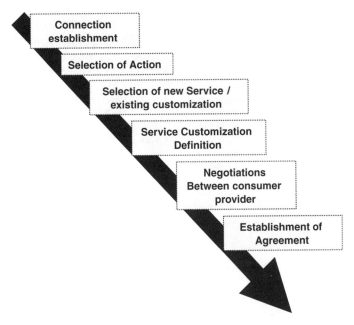

Figure 3.7   Service Customization Component Functionality.

The detailed functionality of the Service Customization component, starting from a consumer contacting the service provider and finishing with the forwarding of the agreed service customizations for this consumer, is presented in Figure 3.7.

A connection is established between the consumer and the provider. Through this connection a negotiation concerning the provider's offered services will take place. A list of available actions will be presented to the consumer (e.g., subscribe to new service, alter customizations, terminate subscription). The consumer must then make a selection using this list. Based on the selection, if deemed necessary, the consumer must provide customizations for the service reflecting his particular needs. The provider processes the consumer's request and either accepts it or rejects it (e.g., because of inability to meet certain requirements) or proposes an alternative set of customizations, close to the requested ones, that can be fulfilled. When agreement is reached, a subscription is established and the required service customizations are forwarded to the Code and Policies Generation Engine in order to be deployed.

The functionality of the Service Customization component is realized by the computational components presented in Figure 3.8.

The Consumer Service Customization GUI is the graphical interface that the consumer uses to communicate with the provider. This interface is responsible for presenting to the consumer all the choices offered by the provider, forwarding the consumer's selections and receiving and displaying the provider's messages. The

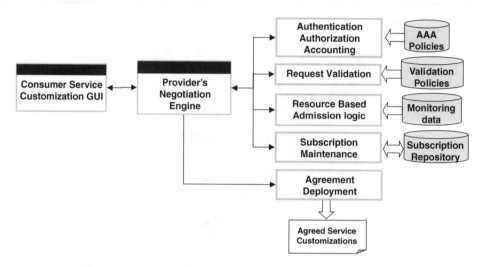

Figure 3.8   Service Customization Component Architecture.

Consumer Service Customization GUI must be dynamic and flexible enough to be able to automatically depict the changes of the provider's offerings (e.g., new services, enhancements on existing services). The Provider's Negotiation Engine is responsible for servicing the consumer's requests coming from the Consumer Service Customization GUI. To service these requests several underlying computational components must be activated, each one performing its designated task and reporting back the results to the Provider's Negotiation Engine, which in turn notifies the consumer. The first of the backend components is the Authentication, Authorization, Accounting (AAA) component responsible for checking and verifying the consumer's credentials, authority, and accounting data, based on dynamically set AAA policies. The Request Validation component validates the consumer's request based on general and service-specific rules defined as Validation Policies. The Resource-Based Admission logic component decides whether the system is able to provide the service with the requested customization, based on monitoring data gathered by the Service Assurance component. If the service cannot be provided, then either the request is dropped or a feasible alternative service customization is computed and proposed to the consumer. The Subscription Maintenance component stores, retrieves, updates, and deletes the established subscriptions that activate the agreements with the consumers. The Agreement Deployment component is responsible for forwarding the agreed service customizations to the Code and Policies Generation Engine that will in turn ensure their deployment.

Multiple instances of the Consumer Service Customization GUI may exist, each one offering a different interface for service customization (e.g., web interface, IVR). All these instances are connected to a single Provider's Negotiation Engine, which in turn relies on a series of backend computational components. The

Provider's Negotiation Engine should be multithreaded in order to be able to support multiple simultaneous consumers requests. The backend computational components are configured by accessing data from specialized repositories (policies, monitoring data, etc). Candidate implementation technology for the computational components is the J2EE platform that enables the creation of multitier, dynamically configured applications. For the repositories, database technology such as Oracle can be used.

3.3.3. Code and Policies Generation Engine

The Code and Policies Generation Engine produces and disseminates the general and customized service code, as well as the management policies necessary for the service provisioning.

As shown in Figure 3.9, this component accepts the Service Definition Document from the CAS Authoring component and the Service Customizations from the Service Customization component. Based on these inputs it delivers the Customized Service code, that is the actual code that implements the logic of the service fulfilling the demands of a service subscription. The customized service code is implemented using an appropriate technology that is supported by the execution environment. In addition, the Code and Policies Generation Engine delivers the management policies resulting from the definition epoch, pertinent to the service, and also policies resulting from the customization epoch, pertinent to specific subscriptions of the service. The management system components provision the services based on these policies.

The component dispatches the resulting code and management policies to the corresponding system components, as shown in Figure 3.9.

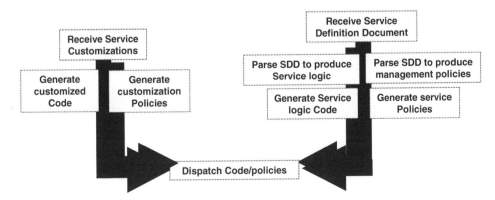

Figure 3.9 Code and Policies Generation Engine Component Functionality.

Once received, the Service Definition Document is parsed to produce the data structures that represent the defined service logic and the derived management policies. These two data structures are utilized for the automatic generation of the service logic code and the service policies, respectively. The data structures, like the Service Definition Document itself, are implementation technology independent, while the code and the policies produced by this component are technologically compatible with the components that implement the CAS provisioning system. When complete, the service logic code is forwarded to the Code Distributor component that will ensure its optimal installation and maintenance, while the service policies are dispatched to the appropriate system components for realizing their necessary configurations that will ensure the provisioning of the new service.

The service customizations, pertinent to the newly established agreement with the consumer, are translated into customized service code and customization policies related to the particular service. The customized service code augments the general service code in order to fulfill the requirements imposed by the agreement with the consumer, while the customization policies configure the system's components in accordance with the new agreement. As before, the customized service code is forwarded to the Code Distributor and the customization policies are dispatched to the appropriate system components (Figure 3.10). The functionality of the Code and Policies Generation Engine component is realized by the computational components presented in the above mentioned figure.

The Service Logic Parser decomposes the Service Definition Document and analyzes all the existing modules pertinent to the construction of the service logic. It produces the Service Logic data structure that is a coherent roadmap for the construction of code that can implement the service logic. The Code Generator component in turn compiles this data structure into technology-specific code. If the code is destined to operate in different execution environments, then a Code Generator must be constructed for each distinct technology, although the Service

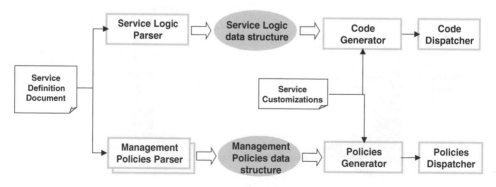

Figure 3.10   Code and Policies Generation Engine Component Architecture.

Logic data structure remains unchanged. The Management Policies Parser is responsible for parsing modules pertinent to specific components or specific aspects of service management and producing the Management Policies data structure with the derived policies. This structure follows the Policy Core Information Model and is technology independent. The Policies Generator component materializes the Management Policies data structure into implemented policies specific for each component of the provisioning system. Finally, the produced code is forwarded to the appropriate installation mechanisms by the Code Dispatcher component, and the produced policies are forwarded to the corresponding management components by the Policies Dispatcher. The Code Generator and the Policies Generator are also responsible for compiling the service customizations into the corresponding customized service code and the customization policies.

The service logic and management policies parsers could be based on available XML parsing libraries and would be sufficiently generic to process any service definition document that conforms to the modeling of CAS, as defined in previous sections. The Management Policy Parser needs to be sufficiently flexible to support the production of new management policies that enhance the system's functionality. Therefore, it must be constructed by multiple classes, each one of them responsible for deriving policies from a set of Realization modules.

3.4. Service Management

As previously mentioned, service management starts when the service code is ready to be deployed in a distributed execution environment. Service management has to play an important role in materializing QoS in Internet [38,39]. Service assurance, one of the phases of service management, has the goal to keep the SLA between users and providers [36] even when the service is supported through different administrative domains [37]. However the set of functions to be considered is heterogeneous and, in general, it is not easy to have closed solutions applicable to any environment. Therefore, some authors structure this functionality as a service itself [16]. In that respect, it is interesting to consider the methodologies to analyze the interactions taking place in the environment of services and in particular the relations between customer and provider of a service in the operation phase [17,18].

One of today's major requirements for service management is the attribute of autonomy. In order to accomplish that goal, it is necessary to have a clear view of the dependencies between the high-level application and its constituent parts and resources [19]. Abstract service models, tools, and languages [42,43,49] are used to address this issue. The problem of system monitoring and the corresponding approaches to make it feasible does not have to be underestimated [40]. In summary, service management needs the use of the most advanced technologies to cope with its different challenges. In this sense, the use of conventional modeling tools like UML [21] or XML [20], Web-based technologies [19] and also policy-based management solutions have been proposed [45–47].

Service management frameworks have evolved from their counterpart at the network level to more and more distributed solutions [35,48]. Today, in many cases, service management solutions requires the existence of distributed service platforms [41]. Nevertheless, it is interesting to have in mind as a reference the most prominent initiatives that have influenced the development of the current discipline. Among these we can mention the efforts of the TINA consortium, the Telemanagement Forum (TMF), and the Distributed Management Task Force (DMTF). The TINA service architecture [23] introduces a set of concepts, principles, rules, and guidelines for constructing, deploying, operating, and withdrawing TINA services. The TINA definition of service management is mainly based on the concepts introduced by network and systems management of TMN/ OSI. The Telecom Operations Map (TOM) [32] introduced by TMF focuses on the end-to-end automation of communications operations processes. The core of TOM is a process framework that postulates a set of business processes that are typically necessary for service providers to plan, deploy, and operate their services. The Common Information Model (CIM) [33] of the DMTF introduces a management information model to integrate the information models of existing management architectures. The Core Model [34] gives a formal definition of a service and allows hierarchical and modular composition of services consisting of other services.

3.4.1. Code Distributor

The Code Distributor is intended to distribute the general service code and the customized service code at selected points in the infrastructure, often for intermediate storage, as well as to carry out the required software maintenance processes. In addition, this component must notify the Code Execution Controller with the optimum URL of the distributed code that implements a given customized service.

As shown in Figure 3.4, the input to the Code Distributor is the Customized Service code that is received from the Code and Policies Generation Engine while the output is twofold. First, it carries out a set of actions that perform the optimal distribution of the customized service code within the network and actions that ensure the reliable maintenance of the distributed code. We call these actions *Code Installation and Maintenance*. Second, the *Service Code URL* answers requests from the Code Execution Controller with the location of the customized service. The returned URL is the most convenient in terms of accessibility by the execution environment.

The installation function of the Code Distributor component starts when the newly created code is available and finishes with the installation of this code into the selected storage points of the underlying infrastructure. The criteria for this selection are provided by management policies acquired during the service definition epoch, with the goal of achieving the optimum code distribution. General policies, established by overall system administration decisions, may also be employed. The maintenance function is always active and ensures that the installed code is stored consistently within the infrastructure that is, keeping a record of code versions, adding new code, removing obsolete code, relocating code for optimizing its distribution, and at all times keeping an updated list of the URLs of the available code. The URL selection mechanism is activated by code URL requests from the Code Execution Controller. These requests ask for the most convenient URL of a specific customized service given a list of the intended execution points for this service. The Code Distributor is responsible for determining the optimum URL, based on the installed code, the URL list, and the management policies defined to affect this selection. These policies and the policies influencing the code installation are acquired during the service definition epoch or determined by global system administration decisions.

The functionality of the Code Distributor component can be realized by the computational components presented in Figure 3.11. The Code Installer component employs a listener for code installation requests, which receives the target code. It maintains a catalog of all the available Code Storage points within the infrastructure. Based on this catalog, and on the management policies, it generates a list of target storage points for the code in question. This list is forwarded to a mechanism that

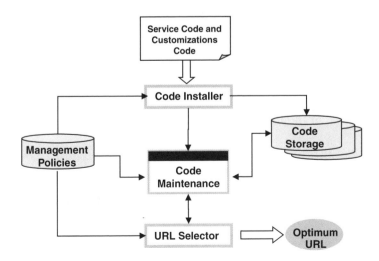

Figure 3.11 Code Distributor Component Architecture.

connects with the code storage points and installs the code. The Code Maintenance component is notified by the Code Installer component for every new installation and keeps an updated list of the code distribution. Removal, update, and redistribution requests are also served by this component. The URL Selector offers an interface for receiving service code URL requests from the Code Execution Controller and answering with the optimum URLs. It also implements the logic that computes the optimum URLs utilizing the pertinent management policies and the code distribution list provided by the Code Maintenance module.

From the implementation point of view the Code Distributor is a centralized component. It is kept up to date with the available code storage points within the infrastructure by the system administrator. A GUI must be provided for this update, this GUI could provide monitoring and human-driven management facilities for the installed code. The Code Storage mechanism could be implemented as a file structure, while the code installation could be achieved using FTP commands. An LDAP-based storage and retrieval system could be considered as an alternative solution.

3.4.2. Code Execution Controller

The Code Execution Controller is intended to derive the appropriate customized services (one or more) to be activated and the terms of their activation. Also, it must deduce the nodes (one or more) hosting the Execution Environments where the code of these services could execute and, finally, it must compose and forward the activation request messages to the corresponding Execution Environments.

As shown in Figure 3.4, the Code Execution Controller interacts with several functional components. Specifically, it receives the activation trigger from the Invocation Service Listeners, the service code URL from the Code Distributor, and the policies from the Service Assurance component. These inputs are processed by the Code Execution Controller, which then issues a service activation request consisting of a message dictating the execution of specific customized service code, at specific node(s) of the Execution Environment and according to specific terms or arguments.

Figure 3.12 depicts the functional processes to be executed by the Code Execution Controller, starting with the processing of an incoming invocation message originating from an Invocation Service Listener and finishing with the forwarding of an activation request message to the Execution Environment.

The invocation trigger message is received and parsed, and the information it conveys is extracted in the form of parameters. These parameters, combined with a set of management policies, will result in the compilation of a list (one or none is also an option) containing the trigger and run time arguments of the services. The

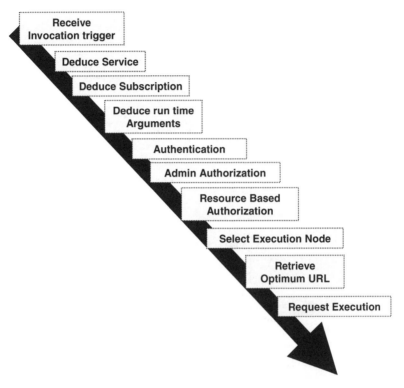

Figure 3.12   Code Execution Controller Component Functionality.

next step is the authentication and authorization for the use of the resulted services. Again the deduced parameters with the list of the resulted services are combined with a set of policies pertinent to the AA functionality, automatically created. The resource-based authorization that follows the administrative authorization will rely on policies set by the Service Assurance component, as a result of its proactive assurance enforcement functionality. Following successful authorization, the execution point (one or more) for each service must be determined. Again this is the result of policies pertinent to execution distribution and policies imposed by proactive assurance enforcement. Once the service execution points are determined, the URLs of the service code are requested to the Code Distributor component. The request includes the execution points of each service, which are necessary for the Code Distributor to reply with the optimum URL for each case. The final step is the composition of activation request messages, targeted at selected points of the Execution Environment, for each service. Each message must include the URL and the runtime arguments of the service code to be executed.

The functional decomposition of the Code Execution Controller is presented in Figure 3.13. The Invocation Triggers Listener receives invocation triggers from

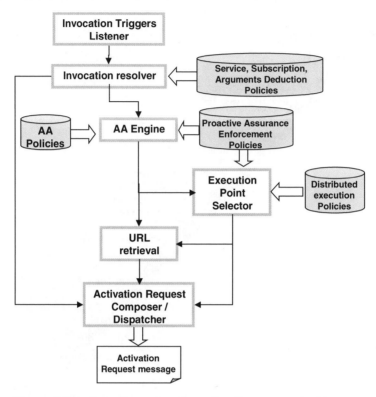

Figure 3.13   Code Execution Controller Component Architecture.

various Invocation Listeners, executes a parsing process, and produces a list of the conveyed invocation parameters. The invocation trigger messages are uniform because the Invocation Listeners undertake the task of hiding the diversity of the various triggering mechanisms (signaling, infrastructure alarms). The Invocation Resolver component receives the extracted invocation parameters and composes a list of customized services (corresponding to established subscriptions with the consumers) by utilizing the relevant policies. These customized services are designed to be activated by this trigger. The list of customized services and the invocation parameters are forwarded to the AA Engine, which authenticates and authorizes the use of these services based on relevant policies. The Execution Point Selector is responsible for selecting the execution points of the services approved for use by the AA Engine, for each specific invocation. The URL Retrieval component implements the connection with the Code Distributor and the retrieval of the services' URLs for each of their selected execution points. The Activation Request Composer/Dispatcher component composes the activation request messages to be dispatched to the execution environment. For each distinct customized service a separate message is composed. This message, shipped to the selected nodes of the execution environment, causes service execution.

The Code Execution Controller could be implemented as a centralized component serving all Invocation Service Listeners and all services. Multiple instances may be configured in order to optimize performance, increase scalability, and improve resilience of the system. Each of these instances would be responsible for serving a number of Invocation Listener components and/or a number of services. The Invocation Listeners must be configured in this case with the necessary policies for selecting the appropriate Code Execution Controller.

3.4.3. Invocation Service Listener

The purpose of the Invocation Service Listener is to capture triggers arriving in the form of various protocol signals or infrastructure alarms, and to encapsulate the contained information in activation trigger messages, which are then forwarded to the Code Execution Controller (see Figure 3.4).

Figure 3.14 presents the functionality of the Invocation Service Listener component in sequence. The process starts with an incoming trigger in the form of a protocol signal or an infrastructure alarm, and finishes with the forwarding of an activation trigger message to the appropriate Code Execution Controller.

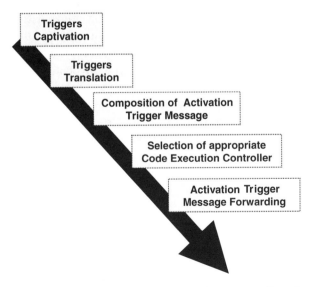

Figure 3.14 Invocation Service Listener Component Functionality.

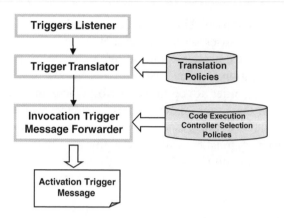

Figure 3.15   Invocation Service Listener Component Architecture.

Each Invocation Service Listener is programmed to recognize only a specific type of trigger (e.g., SIP messages). The captured triggers are translated and the relevant invocation information they convey is deduced with the help of translation policies. These policies are set by the system administrator and are intended to configure the Invocation Listeners in a flexible and dynamic manner, thus making their functionality more generic. The extracted invocation information is encapsulated as parameters in an activation trigger message, which is forwarded to a Code Execution Controller, selected according to policies set during the configuration of this component. The functionality of the Invocation Service Listener component is realized by the computational components presented in Figure 3.15.

The Triggers Listener is responsible for capturing relevant triggers. It is either connected to a signaling termination server or to an infrastructure capability. The Trigger Translator translates the captured messages with the help of a set of translation policies and assembles the conveyed invocation information that is subsequently compiled into an activation trigger message. The invocation information is formalized within the message as a set of parameters, each one having a name and a value. The Invocation Trigger Message Forwarder is responsible for selecting the appropriate Code Execution Controller, based on the relevant policies, and transmitting the message to it. If only one Code Execution Controller is configured, then the Invocation Trigger Message Forwarder does not make a selection, but simply forwards the message directly to the Code Execution Controller.

From an implementation point of view, an Invocation Service Listener must be instantiated for each of the supported trigger types. For performance and scalability reasons a trigger type may be served by several Invocation Service Listeners, each located with a view to reducing trigger transmission delays, reducing network traffic, and balancing server loads. The triggers may take the form of specific protocol signals, or infrastructure alarms produced by context computational objects especially constructed to aggregate required context info into an alarm notification. The signals can be captured by connecting to signaling termination servers, while the infrastructure alarms can be captured by connecting to appropriate context brokers.

3.4.4. Service Assurance

Traditionally, network service providers have offered their subscribers a variety of service quality guarantees, most of which are contractual. Nowadays, new and more demanding applications coexist in today's networks. Each of these applications (e.g., real-time audio or FTP transfers) has different traffic requirements such as bandwidth, maximum delay, and jitter, which must be satisfied in order to ensure adequate service performance. Furthermore, QoS is perceived by the end-user in other ways such as service accessibility, service retainability, or service integrity,[1] which can be defined as:

- Service accessibility performance: perception or measurement of the time taken to access the service, for instance, how fast the user connects to the network.
- Service retainability performance: perception or measurement of how well a service is maintained throughout the usage period without any abnormal interruption or operational outages.
- Service integrity performance: perception or measurement of how well a service is maintained from an end-to-end point of view throughout the usage period, for instance, if a user is receiving the expected network Quality of Service.

Service providers require management systems that can retrieve, calculate, and present quality of service guarantees such that both the provider and the consumer can be assured that service quality meets agreed levels. In this respect, the management system's purpose is to monitor Service Level Agreements (SLA) as defined between the provider and the consumer, and to react to service quality violations. Service assurance automatically detects and corrects network and service problems during service life time, in order to comply with SLA, using policies to achieve this in an efficient manner.

As depicted in Figure 3.4 the Service Assurance component interacts with other functional modules, for example accepting monitoring data from the network Execution Environment. It also sends configuration parameters and individual code to the Execution Environment in order to optimize the performance of the executing customized service code. Finally, the Service Assurance also interacts

---

[1]Other performance areas would be service support, operability, or security, but these will not be dealt with here.

with the Code Execution Controller sending policies that will best configure the Code Execution Controller to optimize processing of invocation triggers. These policies will affect invocation admission and execution node selection processes.

The management system must identify problems affecting network or service provision performance and undertake any actions required in order to achieve QoS levels agreed with end users. The policy-based features of the system contribute to make CAS assurance a fully automated process, allowing high levels of flexibility and dynamic configuration of service monitoring and problem-solving actions.

Two main aspects can be identified in service assurance: service monitoring and service control actions. These are described below.

3.4.4.1. Service Monitoring

We can identify three main functions in service monitoring, as follows:
Performance monitoring: This function is intended to supervise network and service performance. It takes into account four basic service performance parameters for all services:

- Utilization
- Performance
- Reliability
- Congestion

Those four parameters are general but are managed in a different way for each service. In particular, the system considers the use of specific performance parameters for each service.
Threshold data monitoring: The main goal of this function is to provide the mechanisms to configure thresholds applied to service-effecting measurements, and to report the threshold violations to the interested parties via alarm mechanisms.
Fault data monitoring: This function provides the means to report the occurrence and clearance of network or service failures. Failures at the network or service levels are reported using notifications and alarms.

3.4.4.2. Service Control Actions

The management system must take actions in order to maintain the QoS agreed in the SLAs between the provider and the client. We can differentiate two main categories of actions at this point: reactive actions, which are executed once an SLA is violated and preventive actions, which are performed to prevent SLA violations from occurring.

As a trivial example, consider a link between two nodes, and assume that at a certain point in time, the traffic over the link starts to increase in a specific manner. A reactive action would be to set up an auxiliary link between both nodes, as soon as the offered traffic exceeds the link capacity. Clearly, during setup time of this new link, part of the traffic might be lost. On the other hand, a proactive measure would anticipate the situation by studying the evolution of traffic increase and activate the additional link before traffic is lost. Although more efficient, proactive actions require a more complex behavior algorithm.

Another important issue to be considered is the definition of different levels of priority in the actions to be taken and therefore different alarm levels. As an example, the failure of a VPN connection used to provide a service will require immediate action, but if a quality parameter descends slightly without violating the SLA, there is no need to react immediately.

We can categorize the possible management actions as monitoring state actions and monitoring entity actions, as follows.

Monitoring state actions   do not explicitly resolve SLA violations, but help to identify and avoid them. Actions such as changing threshold levels, activating/ deactivating alarms, or adding new monitoring jobs, all fit into this category. For example, when an alarm is triggered, the monitoring system may automatically change the testing period or the alarm threshold for the same or other observed attributes, or perform additional processing to complement existing monitoring actions.

Monitoring entity actions   are taken before or after (proactive vs. reactive) the SLAs are violated, in the first case to avoid the SLAs violation, and in the second case to restore a normal service performance. Examples of these actions are activating auxiliary links when bandwidth problems arise or rejecting new connections if a degradation of service performance is detected.

A detailed functional architecture of a Service Assurance component is presented in Figure 3.16, which shows the two main components, namely the Service Performance Asserter and the Monitoring Component, which are in turn decomposed into other modules as described hereafter.

The Fault Monitor performs functions and provides capabilities to maintain, manage, and report fault data to other management systems. It enables a management application such as the Service Quality Manager or the Assurance Manager to receive, correlate, and manage faults produced by/for different managed objects.

The Performance Monitor performs functions and provides capabilities to acquire, monitor, and report performance data for a set of managed objects. It also provides capabilities to set performance thresholds, monitor, and report threshold crossings.

The Service Quality Manager is responsible for monitoring and reporting aggregated service quality and performance data obtained from different raw data sources. It is responsible for configuring the different monitoring jobs (change thresholds, etc.), if necessary.

The Assurance Manager consists of a set of functions that manage the life cycle of network and service problems. It decides whether changes to network components must be undertaken, based on the information that the Fault Monitor and the SLA

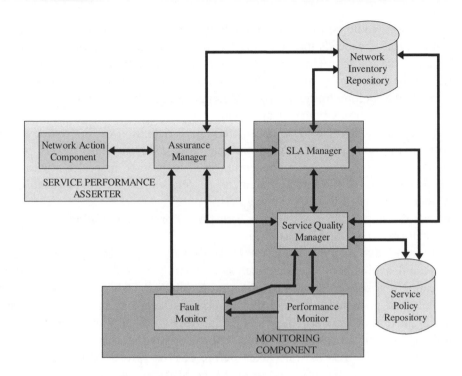

SERVICE ASSURANCE FUNCTIONAL ARCHITECTURE

Figure 3.16   Service Assurance Component Architecture.

Manager submit. It can also send a request to the Service Quality Manager to adjust parameters of the monitoring jobs.

The SLA Manager is responsible for the correlation of service quality and performance data with specific customer service instances to identify any SLA violations. If any violation occurs, he can inform the Assurance Manager and let him decide whether actions to reallocate network resources are necessary (through the Network Action Component), if configuration of monitoring jobs is needed (through the Service Quality Manager), or both.

The Network Action Component makes corrective or preemptive adjustments of network or service resources.

The Network Inventory Repository maintains the configuration of the network objects and their association with each service instance. The Service Policy Repository maintains the configuration of the policies and their association with each service instance.

The Network Action Component is responsible for taking actions to reallocate network or service resources. Actions previously referred to as Monitoring State Actions are controlled and caused by interactions between the Service Quality

Manager component and the various monitoring components (Performance Monitor and Fault Monitor).

From an implementation point of view, the Service Assurance component can be conceived as partially distributed. Therefore, we distinguish between its distributed components and its centralized ones:

- Distributed components: each node has its own instance of each of these components. Service Quality Manager, Performance Monitor, and Network Action Component are envisioned as distributed components.
- Centralized components: these components have a centralized nature, although there might be multiple instances of them over a large network. Assurance Manager, SLA Manager, and Fault Monitoring are thought of as centralized components, as they need information from different nodes and that causes a considerable complexity in decision making. For example, the SLA Manager will probably need performance measures from different nodes to decide whether a SLA has been violated or not.

In order to communicate with different nodes for assurance purposes, the Service Quality Manager has external APIs in addition to the APIs offered to the internal assurance components on the same node. The external APIs allow remote nodes to access and process the monitoring information available at the local node. For example, assume that node A has two possible routes to reach node D. These routes are through node B and C. At any given point in time, node A may check the status of the interfaces of node B and C (congestion, failure, etc.) in order to decide which is the best route to reach node D and, on this basis, perhaps alter its routing table.

3.5. Conclusions

This chapter describes the service life cycle of Context-Aware Services. The definition, customization, invocation, and operation epochs are described and their functional components identified. Finally, the functional components, which are constituted into a full functional architecture of the system that enable the whole service life cycle system, are described in detail.

References

1. ETSI Technical Report ETR 323: 'Service life cycle reference model for services supported by an IN,' December 1996.
2. ETSI Technical Report ETR 137: 'Intelligent Network (IN); Service and feature interaction: Service creation aspects, service management aspects and service execution aspects.' 1995.

3. W3C: Web Service Management: Service Life cycle. http://www.w3.org/TR/2004/NOTE-wslc-20040211/.
4. Gervais MP, Diagne A. Enhancing telecommunications service engineering with mobile agent technology and formal methods. *IEEE Communications Magazine* **36**(7): 1998; 38–43.
5. Reynolds PL, Sanders PW. Integrated services engineering, Fifth IEE Conference on Telecommunications, 26–29 March 1995, 297–301.
6. Hayes MJ. Telecommunications service engineering. Intelligent Networks: IEE Tutorial Seminar on Advanced Services and their Management, 11 May 1994 pp 311–313.
7. Niemela E, Kalaoja J, Lago P. Toward an architectural knowledge base for wireless service engineering. *IEEE Transactions on Software Engineering* **31**(5): 2005; 361–379.
8. Kirda E, Kerer C, Kruegel C, Kurmanowytsch R. Web service engineering with DIWE. In *Proceedings of* 29*th Euromicro Conference*, 1–6 September 2003, p 283–290.
9. Dobson J. Issues for service engineering. In *Proceedings of First International Workshop on Distributed and Networked Environments*, Services in 27–28 June 1994, pp 4–10.
10. Eastman J, Fuller I, Hirschfeld R. Service life cycle in a distributed computing environment. *Telecommunications Information Networking Architecture Conference Proceedings*, TINA '99 12–15 April 1999, pp 183–184.
11. Chen YP, Li ZZ, Jin QX. A whole life cycle model to dynamic composed web services. In *Proceedings of 2005 International Conference on Machine Learning and Cybernetics*, 2005; Volume 2, 18–21 August 2005, pp 1047–1052.
12. Hasselmeyer P. An infrastructure for the management of dynamic service networks. *IEEE Communications Magazine*. 2003; **41**(4) 120–126.
13. Huang Y, Kumaran S, Chung J-Y. A service management framework for service-oriented enterprises. In *Proceedings of IEEE International Conference on e-Commerce Technology*, CEC 2004, 6–9 July 2004, pp 181–186.
14. Kar G, Keller A, Calo S. Managing application services over service provider networks: architecture and dependency analysis. NOMS 2000. *IEEE/IFIP Network Operations and Management Symposium*, 10–14 April 2000, p 61–74.
15. Luling R. Managing large scale broadband multimedia services on distributed media servers. *IEEE International Conference on Multimedia Computing and Systems* 1999; **1**: 320–325.
16. Strick L, Wittig M, Paschke S, Meinkohn J. Development of IBC service management services. *IEEE Network Operations and Management Symposium* 1996; **2**: 424–433.
17. Garschhammer M, Hauck R, Hegering H-G, Kempter B, Radisic L, Roelle H, Schmidt H. A case-driven methodology for applying the MNM service model. NOMS 2002, IEEE/IFIP Network Operations and Management Symposium, 15–19 April 2002, pp 697–710.
18. Garschhammer M, Hauck R, Hegering H-G, Kempter B, Radisic I, Rolle H, Schmidt H, Langer M, Nerb M. Towards generic service management concepts a service model based approach. IEEE/IFIP International Symposium on Integrated Network Management Proceedings, 14–18 May 2001, pp 719–732.
19. Boutaba R, El Guemioui K, Dini P. An outlook on intranet management. *IEEE Communications Magazine* 1997; **35**(10): 92–99.

20. Alipio P, Lima S, Carvalho P. XML service level specification and validation. ISCC 2005, 10th IEEE Symposium on Computers and Communications, 27–30 June 2005, pp 975–980.
21. Adamopoulos DX, Pavlou G, Papandreou CA. A UML based methodology for the creation of TINA compatible telecommunications services. ISCC 2000. Fifth IEEE Symposium on Computers and Communications, 3–6 July 2000, pp 653–658.
22. Adamopoulos DX, Pavlou G, Papandreou CA. Advanced service creation using distributed object technology. IEEE Communications Magazine 2002; **40**(3): 146–154.
23. TINA-c: Service Architecture Version 5.0. TINA Baseline, TINA Consortium, June 1997.
24. Rana S, Fisher MA, Egelhaaf C. Implementation and interoperability experiences with TINA service management specifications. Telecommunications Information Networking Architecture Conference Proceedings, 1999, TINA '99.
25. Manione R, Renditore P. A 'TINA light' service architecture for the Internet-telecom scenario. Telecommunications Information Networking Architecture Conference Proceedings TINA '99, 12–15 April 1999, pp 24–32.
26. Pavon J, Tomas J, Bardout Y, Hauw L-H. CORBA for network and service management in the TINA framework. *IEEE Communications Magazine* 1998; **36**(3): 72–79.
27. Park HS, Choi O-H, Baik D-K. CORBA based approach to the development of an advanced architecture in TINA service management system. 12th International Workshop on Database and Expert Systems Applications, 2001, 3–7 September 2001, pp 175–179.
28. Lin Y-D, Lin Y-T, Chen P-N, Choy MM. Broadband service creation and operations. *IEEE Communications Magazine* 1997; **35**(12): 116–124.
29. Kirda E, Jazayeri M, Kerer C, Schranz M. Experiences in engineering flexible Web services. *IEEE Multimedia* 2001; **8**(1) 58–65.
30. Cortese G, Fiutem R, Cremonese P, D'antonio S, Esposito M, Romano SP, Diaconescu A. Cadenus: creation and deployment of end-user services in premium IP networks. *IEEE Communications Magazine* 2003; **41**(1): 54–60.
31. Prodan R, Fahringer T. From Web services to OGSA: Experiences in implementing an OGSA-based grid application. Fourth International Workshop on Grid Computing, 17 November. 2003, pp 2–9.
32. Telecom Operations Map. Approved Version 2.1 GB910, TeleManagement Forum, March 2000.
33. Common Information Model (CIM) Core Model. White paper, Desktop Management Task Force, August 1998.
34. Common Information Model (CIM) Specification Version 2.2. Specification, June 1999.
35. Dias B, Santos A, Boavida F.Internet network services management framework. ICON 2002. 10th IEEE International Conference on Networks, 27–30 August 2002, pp 361–368.
36. Chakravorty R, Pratt I, Crowcroft J, D'Arienzo M. Dynamic SLA-based QoS control for third generation wireless networks: the CADENUS extension. ICC '03. *IEEE International Conference on Communications* 2003; **2**: 938–943.
37. Baek J-W, Park J-T, Seo D-i. End-to-end Internet/intranet service management in multi-domain environment using SLA concept. NOMS 2000. IEEE/IFIP Network Operations and Management Symposium, 10–14 April 2000, pp 989–990.

38. Mykoniati E, Charalampous C, Georgatsos P, Damilatis T, Goderis D, Trimintzios P, Pavlou G, Griffin D. Admission control for providing QoS in DiffServ IP networks: The TEQUILA approach. *IEEE Communications Magazine* 2003; **41**(1): 38–44.
39. Giammarco C, Malick K, Morreale P. Wireless quality of service assurance for network survivability. MILCOM 1999. *IEEE Military Communications* 1999; **2**: 893–896.
40. Asgari A, Egan R, Trimintzios P, Pavlou G. Scalable monitoring support for resource management and service assurance. *IEEE Network* 2004; **18**(6): 6–18.
41. Adamopoulos DX, Pavlou G, Papandreou CA. Development of new telecommunications services in distributed platforms: A structured approach. ICC 2000. IEEE International Conference on Communications, 2000, Vol. 1, 18–22 June 2000, pp 222–226.
42. Gopal R. Unifying network configuration and service assurance with a service modeling language. NOMS 2002, IEEE/IFIP Network Operations and Management Symposium, 15–19 April 2002, pp 711–725.
43. Choi S-H, Ha J-H, Song J-G. Building a service assurance system in KT. NOMS 2004. IEEE/IFIP Network Operations and Management Symposium, Vol. 2, 19–23 April 2004, pp 73–86.
44. Panagiotakis S, Alonistioti A. Intelligent service mediation for supporting advanced location and mobility-aware service provisioning in reconfigurable mobile networks. *IEEE Wireless Communications* 2002; **9**(5): 28–38.
45. Flegkas P, Trimintzios P, Pavlou G. A policy-based quality of service management system for IP DiffServ networks. *IEEE Network* 2002; **16**(2): 50–56.
46. Hong L, Dong B, Wei D. A policy-based solution for management of enhanced network services. TENCON '02. 2002 IEEE Region 10 Conference on Computers, Communications, Control and Power Engineering, Vol. 3, 28–31 October 2002, pp 1684–1687.
47. Badr N, Taleb-Bendiab A, Reilly D. Policy-based autonomic control service. POLICY 2004. Fifth IEEE International Workshop on Policies for Distributed Systems and Networks, 7–9 June 2004, pp 99–102.
48. Rayes A, Sage K. Integrated management architecture for IP-based networks. *IEEE Communications Magazine* 2000; **38**(4): 48–53.
49. Adamopoulos DX, Papandreou CA. Object-oriented development of telematic services. ISCC '98 Third IEEE Symposium on Computers and Communications, 30 June–2 July 1998, pp 276–280.

4

Context-Aware Services and the Network Layer

This chapter is an analysis of the requirements posed by context-aware services on the network layer. It starts with the motivation for a special support by the network layer followed by a short review of the current state-of-the-art and service-aware networks. Network support for services is realized via openness through Application Programming Interfaces (APIs). We describe the requirements for collecting network information which is a special part of the overall context information described in Chapter 2 which we term—'Network Context Information.' We also describe the requirements for the capability of the service to interact with the network layer by changing network configuration for example.

4.1. Network Layer Requirements for Context-Aware Services

In the IP world, the network layer provides end-to-end functionality based on a very simple forwarding mechanism and more complex distributed routing protocols. This simple unreliable basic packet forwarding service combining with the very complex transport protocol (TCP), which is executed only at the connection endpoints, provides a reliable adaptive end-to-end transport which is the basic building block of almost all reliable data services. However, this paradigm of simple network and sophisticated routing and endpoints cannot support modern services that require QoS support in a scalable cost-efficient way.

Modern CASs that offer voice over IP networks must provide the needed QoS (e. g. bounded end-to-end delay). Services that provide streaming capabilities or multimedia transport must also consider various network parameters such as available bandwidth, congestion, and delay. In general, the information that is needed by services is information that describes the state of the network from several aspects like load, availability, best routing, etc. This information,

Fast and Efficient Context-Aware Services   Danny Raz, Arto Tapani Juhola,
Joan Serrat-Fernandez, Alex Galis   © 2006 John Wiley & Sons, Ltd

which we term Network Context Information, is used by services to allow them to deliver the required service behavior to users and to improve their performance.

The nature of the Network Context Information is that it is composed of different pieces of information, residing in different network locations. For example, the route of a path in an IP network is composed of the entries in the forwarding tables of the routers along the path, and it is not available in any centralized entity. The same holds for a list of congested links, over-utilized resources, etc.

There are three critical aspects in the ability to provide the needed network context information to the networked services. The first one is the ability to extract pieces of the needed local information, that is, the ability to get information from network elements regarding their local state and the state of the network. The second step is to create a general network layer view from the local pieces of information. The third critical step is the delivery of the needed information to the different services and, in particular, to the different parts of the service logic, which might be executed in a distributed way as described in the previous chapter.

Taking into account the properties of and requirements underlying the provisioning of Network Context Information, one of the main obstacles is the efficient collection of this information from the network and the ability to process and deliver it to the requesting overlay network as required.

This is a complicated task; it needs to be done 'on request' basis, since we only need to monitor and collect information that is requested by an information client (a context-aware service). This means that we need to allow for the following steps to happen (a) to define the information that clients need (b) after checking that information can indeed be given according to the current policies and abilities of the network, we should be able to employ mechanisms to collect the needed information, and (c) to send it back to the appropriate client. Of course, since all monitoring information is delivered over the network, we need to reduce overhead traffic as much as possible as it competes for the same bandwidth with customer traffic.

Acquiring the needed network information is not enough. In many cases the service needs to take action as to change network configuration in order to be able to deliver the service, or in order to make it more efficient. To that end, there is a need for a mechanism that allows the service to take actual actions with respect to the network. Such an action could be a reconfiguration of parameters in one or more network elements, or setting up filters blocking certain packets in different network locations, etc.

The ability to collect, process, and disseminate network context information, together with the ability to actually change the network behavior by taking actions, is a key ingredient in the ability to design and implement complex context-aware services in a scalable and efficient way.

4.2. Current State of Service-Aware Networks and Open Network Interfaces

Internet has been founded on a basic architectural premise: a simple network service is used as a universal means to interconnect intelligent end systems. The end-to-end argument has served to maintain this simplicity by pushing complexity, where needed, into the endpoints, allowing the Internet to reach an impressive scale in terms of inter-connected devices. However, while the scale has not yet reached its limits, the growth in functionality—the ability of the global system to adapt to new functional requirements and new services and applications—has slowed with time. In addition, the ever-increasing demands of applications and network services, and the envisaged future generations of Internet are now faced with the relative inflexibility of current telecommunication infrastructures.

Introducing openness through APIs into networks is one way to increase the needed flexibility and support for services. Such networks are called service-aware networks. Service-aware networks are better supporting services in terms of resource management, flexibility, programmability, protocol adaptation, reusability, scalability, reliability, and security [7,10,11]. A service-aware network supports adaptation for all applications and services residing on top. The adaptation process takes place in the middleware [4,14]. Services and applications are able to monitor events in the underlying network and adapt their internal behavior by either changing specific components' behavior or dynamically reconfiguring the network [26].

The most well known framework realizing today the concept of open network interfaces is the Parlay/OSA initiative [18]. Built on concepts previously established in the TINA architecture [25], it consists of a set of APIs that enable third parties and network operators to create new applications and services, which have real-time control of network resources. The Parlay framework provides the resource location, authentication, and authorization functions required for external applications to gain access to network-based Parlay services. Parlay services, such as the generic call control service (GCCS), offer network capabilities to applications, enabling them, for example, to dynamically route and reroute media streams.

From a more general point of view, one significant approach towards service-aware networks is based on network overlays [1,2,6,19–21]. Overlay network exhibit a clear border [5,8,9,12,13] below which an overlay abstraction is enabled and above which applications should not have to deal with the underlying 'real' network infrastructure. In overlay networks a set of nodes (servers, end-user equipment, etc.) and virtual links, not directly related to the underlying network topology, are involved in specific applications. The overlay traffic traverses through the overlay nodes and virtual links. Therefore, an overlay network acts as a specialized middleware. Two elements have to be considered: (a) techniques to efficiently map the overlay abstraction to the underlying resources, and (b) the management of the overlay, that is, the mapping policy, configuration and reconfiguration, performance monitoring, etc. One of the benefits of overlays is that they can be customized or optimized for a single service or a service family, thus creating a variety of overlays [7].

A large number of peer-to-peer overlay network designs [10,11,22–24] have been proposed recently, such as CAN [1], Chord [23], Freenet [9], Gnutella [3,17], Pastry [21], Salad, Tapestry, Viceroy, SkipNet. One of the main benefits of peer-to-peer networks is that they enable service components deployment (e.g., content services and content distribution networks) in a flexible, scalable, and decentralized way. A key function that these networks enable is a distributed hash table, which allows data to be uniformly diffused over all the participants in the peer-to-peer system. The basic approach in systems like Pastry and Chord is to diffuse content randomly throughout an overlay in order to obtain uniform, load-balanced, peer-to-peer behavior. In Pastry, Chord, and Tapestry the expected number of hops between any two communicating nodes scales with logN. In SkipNet data is uniformly distributed across a well-defined subset of the nodes in a system, such as all nodes in a single building. Overlays have also appeared over grids [2,15]. A dynamic overlay of active network routers to accomplish scalable time management in a grid environment is presented in Reference [16], while Reference [2] deals with topology awareness.

The separation between services and network infrastructure results in some loss of interworking between the two layers; the programmable network technology offers some remedy to this inefficiency. Active and programmable network technology is described in great details in the next chapter.

4.3. Requirements for Network Context Information Collection and Dissemination

Context-aware services need a constant flow of information about their environment, in order to be able to adapt it. As explained above, an important part of this information is Network Context Information. This information is derived from many different sources spread around the network. A fundamental requirement is to collect raw data from these sources, process it, and as needed disseminate the information to applications (services) at different points of the network. We distinguish between two important aspects of this process. The first step is the ability to access local network level information. Once this ability is available, we need to discuss the processing of the information in the network level and the dissemination of the information to the prospective consumers—that is, the services who need this information.

4.3.1. Access to Local Network Level Information

As explained above, the access to local network information at different network locations is a basic capability that is needed. The type of information that needs to be collected varies depending on the specific service and the specific network device at hand. For a generic network device, relevant identifiers local information at network level can be the associated to all of its interfaces, so that it will be possible to recognize each one and retrieve, in case it is necessary, the associated IP addresses, the offered load, or other parameters of an IP interface. For a routing device, routing information like the next hop associated in the routing table to a given IP destination address may be also of interest. If the local device has the functionality of a WLAN access point, local information will be information allowing the management of service sets (creation, removal, modification), management of VLANs, signal strength, and QoS policies. In case of a GPRS server, the relevant local information can be the connection status of users, their geographical coordinates, or their respective IP addresses.

This type of information is usually accessible either via a standard SNMP agent or using a proprietary CLI. However, in our case we need an efficient channel that will allow access to this information to the service logic; thus, local network level information needs to be abstracted so that upper level applications or entities will not have to worry about dealing with each local devices. This abstraction can be in the form of an information model and a set of APIs based on such an information model. Such an API supports unified queries regarding network level data. Efficiency could be provided by a mediator or a set of mediators that have the necessary intelligence to process information taking into account characteristics like the location of such information, its volatility, the protocol that understands or is best suited to the specific device, and the grouping of several atomic requests into a combined one to reduce the network overheads.

4.3.2. Gathering and Disseminating Global Network Information

Gaining access to local network related is not enough. In many cases the full network picture is a function of different local parameters, each associated with a different network element at a different network location. For example, the delay of a voice over IP session is the sum of the local delay along the path of the connection. We need then to create a network-level picture of the relevant network information and to make it available to the services that need this information.

In order to describe the solution we adopt the provider (some times called producer) consumer approach: each player in the system is either a consumer of network context information – these are the CASs – or a provider of context information. A Context Information Distribution System (CIDS) is the system that

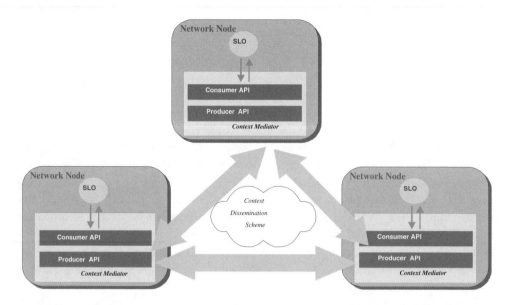

Figure 4.1   Context Information Dissemination System (CIDS).

collects, processes, and distributes this information. Note that while in this chapter we deal with network context information the discussion regarding the CIDS is general and we use the same model in Chapter 7. In such architecture, the CIDS is responsible to bind the consumers and the producers, and to distribute the contexts using well-defined APIs. The key abstraction of these APIs is that data producers generate data by publishing it, data consumers subscribe to data, and it is the business of the CIDS to ensure that context information travels efficiently from publisher to consumer. Thus, we consider the CIDS as a system that has endpoints (called *mediators*) at each relevant location. These endpoints can communicate among themselves and produce the information needed (see Figure 4.1).

The CIDS takes into account any special requirements, such as timely distribution of context to a large set of consumers, common context needs of consumers, asymmetry due to disproportion between the number of producers and consumers, and dissemination of context updates due to the volatile nature of context information. By combining various data delivery techniques, the scalability and performance of the dissemination mechanism can be greatly enhanced. The integrated CIDS consist of context producers, consumers, and mediators. The mediators act as intermediaries between producers and consumers, providing the means for efficient information exchange. The nodes of the CIDS can communicate using a variety of delivery options.

The producers of context information include all the Context Information Sources (CISs) that are attached to specific network nodes and provide raw context information, such as information from network nodes (e.g., local MIB variables), information from different mobile access technology adapters (WLAN, GPRS), information from other wrappers (e.g., weather sensors), and various other information provided by Context Providing Applications (e.g., user agenda). In addition to context producers that provide raw context information, special services called Complex Computation Objects (CCOs) may provide complex information by aggregating elementary context.

The context consumers are the SLOs, that is, the instances of the service logic. During the operation epoch, each SLO issues requests for the acquisition of context information in order to adapt to future context changes and to enforce appropriate actions.

The CIDS mediators, namely the Context Mediators (CMs), act as third-party players between producers and consumers. These mediators accept requests from SLOs, collect context information from producers, and efficiently disseminate the information from producers to consumers, or other CMs, as dictated by a dissemination scheme. In this sense, CMs realize the linkage between producers and consumers. Each CMs provides two different types of API: the Producer APIs and the Consumer APIs. The Producer APIs include the interfaces that enable context producers to publish the information they provide, either raw or complex, so that the CMs will then be able to access it. In this sense, the heterogeneity of the different context sources is hidden and each time a modification of the context sources occurs; only the local Context Mediator needs to be updated with the new interfaces, in order to interact with them. This process is transparent to the SLO. The Consumer APIs enable an SLO to access context information through either 'pull' requests or notification events. In the case of complex context information, the CCOs collect raw context information from the local or remote context sources, aggregate it, and produce complex context information; thus, CCOs represent CISs of complex context. The CCOs are deployed (permanently or on demand) in the EE of the network nodes and are registered with the CMs through the Producer APIs.

4.4. Requirements for Network Level Control

The previous sections describe how the relevant network information is made available to the service logic. The service then uses its logic (in terms of policy) to take actions based on the network context. Some of the needed action may be related to reconfiguring of the network in order to achieve the service goals. In order to do so, the service needs an API that will allow it to perform actions with respect to the network layer.

As in the case of information retrieval, an action can be local, that is, solely performed at a single network element, or global that is, an action that needed to be performed in more than one network location. Examples for local actions could be changing a configuration parameter in the local network device, setting up a filter to block certain traffic, changing QoS priorities, etc. Examples for global actions may include setting up an IP tunnel or changing an OSPF characteristic in an entire region.

While needed for the efficient delivery of CASs, allowing code to perform such action in the network layer presents a serious security risk that must be addressed. Apart from the possibility of malicious code taking over the network, there is also the risk of bugs that may cause the network to malfunction. The problem is more severe when we consider network-level operations where the action needs to be performed simultaneously in more than one location and it needed to be atomic, that is, if it fails in one location it should not be performed in any of the other locations.

Again, a very good way to realize such ability is via a mediator. Such a mediator can abstract the action from the service logic, and can also deal with the security and safety issues in a unified way. Moreover, it can also handle network-level actions and their atomicity by monitoring the status and rolling back the operation when needed.

The Action Mediator realizes all the above by providing an associated API to the SLOs. These APIs provide a uniform way to perform an action at the local network level and at the network level. It is important to note that an efficient implementation of network-level action mediator is by using programmable network techniques (as described in the next chapter). In this way a mediator exists in each relevant network node and performing global action is done by communication among these mediators, where each one is in charge of the actions in the local device. Such an implementation is described in Chapter 7, where the Action Mediator is one of the brokers in the DINA system called the Action Broker.

4.5. Security Considerations

Let us consider a piece of context information, referred to as a Context Item (CI). The CI is offered by a CIS, accessed by several Context-Aware Services (CAS). As the performance of these services may be highly dependent on the value of the CI (depending on the context, service functionality is different), it is important to ensure its security. Security mechanisms are required for context information exchange in order to avoid information tampering and to protect private information from unauthorized access, otherwise the CAS functionality may be seriously compromised. Depending on the sensitivity of the context information or the potential impact on the requesting CAS's behavior, different levels of security can be considered (e.g., different security concerns apply to a CI containing a temperature measure in a public place than to a CI concerning the saturation state of emergency calls to 112).

Three different security mechanisms for securing context information exchanges are considered:

1. Authorization: The CAS requesting the CI must have permission to access it. A CAS can be either authorized or unauthorized.
2. Authentication: There must be a mechanism to certify CIS's identity, ensuring that the information source is trustworthy, and also reducing the possibility of information tampering. Authentication can be achieved through the use of digital signatures.
3. Encryption: The context information is encrypted to protect private information from access or modification during transfer.

There can be different levels of security depending on which of these mechanisms are important to a specific CI. Any combination of mechanisms can be applied. For example:

- $1+2+3$ (Authorization, Authentication, and Encryption): Access to a CIs containing private information that can be expected to be very relevant to the requesting CAS's functionality (e.g., when accessing personal bank account information; we could consider a CAS that reacts in one way or another depending on the available cash in a user's account).
- 2 (Authentication only): Access to a CIS that is commonly expected to provide some public CI in a precise and correct manner; as that CI can strongly influence requesting CASs behavior, authentication allows those CAS who trust in the CIS to obtain reliable information.

Providing the above-mentioned security mechanisms must be responsibility of CMs, as CISs are not assumed to have any knowledge outside the CI that they provide.

4.5.1. Implementation Aspects

The CM's CIS Registration Repository may contain the security level required to access a specific CI at a CIS, recorded during the CIS registration process, to allow CM to provide the desired level of security.

For encryption, a public/private key scheme, using technology such as PGP, may be used. Following this scheme, each CAS or CIS would have a public key and a private key. The public keys would be officially registered in a CM somewhere. The private keys would be accessible in case they were needed.

When a CIS needs to register a CI in a CM one of the required parameters would be a security-level identifier (authorization and/or authentication and/or encryption). Depending on the desired security level, more parameters would be needed. For example, if authorization were required, it would be necessary to specify all the public keys of the SLOs that were authorized to access that CI.

The following scenarios illustrate the use of security mechanisms:

- An SLO asks for a CI: The CM finds in his CIS Registration Repository that authentication mechanisms are offered by the CIS for that Context Item and authenticates the CIS, making use of the CIS's keys.
- An SLO asks for a CI: The CM finds in the CIS Registration Repository that the CI requires encryption. The CM establishes a secure connection between intermediate CMs, as well as between the SLO and the destination CM using the appropriate keys in each case.
- An SLO asks for a CI: The CM finds in the CIS Registration Repository that authorization is required to access the CI. Making use of the appropriate mechanisms, the CM determines if the SLO is authorized to access the CI.

In order to implement these security mechanisms it is fundamental that the distributed nature of context information is considered.

4.6. Conclusions

This chapter describes the special interaction between the services and the network layers that are needed in order to provide scalable efficient context-aware services in future converged networks. We described the state of the art and explained the needed mechanisms that can guarantee access of the services (i.e., the service logic SLO) to the relevant network context information. Furthermore, we described a mechanism (CIDS) that allows collecting processing and disseminating this information to the appropriate services. As mention this mechanism can be used to handle general types of context information, and a more detailed description of it is provided in Chapter 7.

References

1. Abraham I, Awerbuch B, Azar Y, Bartal Y, Malkhi D, Pavlov E. 'A generic scheme for building overlay networks in adversarial scenarios'. *Proceedings of the 17th International Symposium on Parallel and Distributed Processing*, April 2003, Nice, France, pp 22–26.
2. Andersen DG, Balakrishnan H, Kaashoek MF, Morris R. 'Resilient overlay networks'. *Proceedings of 8th ACM SOSP*, Banff, Canada, October 2001.
3. Anonymous. 'Gnut: Console Gnutella Client for Linux and Windows'. http://www.gnutelliums.com/linux_unix/gnut/, 2001.

 4. Coady Y, Kiczales G. Back to the Future: A Retroactive Study of Aspect Evolution in Operating System Code. In *Proceedings of Aspect Oriented Systems Development* AOSD 2003, pp 138–146.

 5. Bassi A, Beck M, Moore T, Plank JS. 'The logistical backbone: Scalable infrastructure for global data grids'. In *Proceedings of Asian Computing Science Conference*, Hanoi, Vietnam, Lecture Notes in Computer Science, Vol. 2550/2002, Springer, December 4–6, pp 1–12.

 6. Braynard R, Kostic D, Rodriguez A, Chase J, Vahdat A. 'Opus: An overlay peer utility service'. In *Proceedings of the 5th International Conference on Open Architectures and Network Programming* (OPENARCH), June 2002.

 7. Campbell AT, De Meer HG, Kounavis ME, Miki K, Vicente J, Villela DA, 'The Genesis Kernel: A virtual network operating system for spawning network architectures'. In *Proceedings. IEEE OPENARCH'99*, New York, March 1999.

 8. Castro, M, *et al.* 'One ring to rule them all: Service discovery and binding in structured peer-to-peer overlay networks'. SIGOPS, France, September 2002.

 9. Clarke I, Sandberg O, Wiley B, Hong TW. 'Freenet: A distributed anonymous information storage and retrieval system'. Workshop on Design Issues in Anonymity and Unobservability, July 2000, pp 311–320.

10. De Meer H, Tutschku K. 'Dynamic operation in peer-to-peer overlays'. Poster Presentation Supplement to the *Proceedings of Fourth Annual International Working Conference on Active Networks*, Zurich, Switzerland, December 4–6, 2002.

11. De Meer H, Tutschku K, Tran-Gia P. 'Dynamic Operation in peer-to-peer overlay networks', Praxis der Informationsverarbeitung und Kommunikation, (PIK Journal), Special Issue on Peer-to-Peer Systems, June 2003.

12. Ghosh A, Fry M, Crowcroft J. 'An Architecture for application layer routing'. *IWAN* 2000; pp 71–86.

13. Wang J.G, Li ZZ, Kou Y N. 'Research and implementation of a scalable secure active network node'. In *Proceedings. of IEEE International. Conference on Machine Learning and Cybernetics*, Vol. 1, November 2002, pp 111–115.

14. Kiczales G, Lamping J, Videira Lopes C, Mendhekar A, Murphy G. Open implementation design guidelines. In *Proceedings of International Conference on Software Engineering*, Boston, M, May, 1997 pp 87–96.

15. Keahey K, *et al.* 'Computational grids in action: The national fusion collaboratory'. *Future Generation Computer Systems* 2002; **18**:(8): 1005–1015.

16. Korpela, E, *et al.* 'SETI@home: Massively distributed computing for SETI'. *Computing in Science and Engineering* 2001; **3**(1): 78–83.

17. Klingberg T, Manfredi R. The Gnutella Protocol Version 0.6 Draft, Gnutella Developer Forum, 2002, http://groups.yahoo.com/group/the_gdf/files/Development/.

18. Moerdijk AJ, Klostermann L. 'Opening the networks with Parlay/OSA: Standards and aspects behind the APIs'. *IEEE Network* 2003; **17**(3): 58–64.

19. Ratnasamy S. 'A scalable content-addressable network'. Ph.D. Thesis, U.C. Berkeley, October 2002.

20. Ratnasamy S, Francis P, Handley M, Karp R, Shenker S. 'A scalable content-addressable network'. In *Proceedings of ACM SIGCOMM*, August 2001.

21. Rowstron A, Druschel P. 'Pastry: Scalable, distributed object location and routing for large-scale peer-to-peer systems'. International Conference on Distributed Systems Platforms (Middleware), Heidelberg, Germany, November. 2001, pp 329–350.

22. Schollmeier R. 'A definition of peer-to-peer networking for the classification of Peer-to-Peer architectures and applications', First International Conference. on Peer-to-Peer Computing (P2P2001), LinkÎping, Sweden, 2001.

23. Stoica I, *et al.* 'Chord: A scalable peer-to-peer lookup service for internet applications'. ACM SIGCOMM '01, San Diego, CA, September 2001.

24. Stoica I, Morris R, Karger D, Kaashoek MF, Balakrishnan H. 'Chord: A scalable peer-to-peer lookup service for internet applications'. In *Proceedings of ACM SIGCOMM*, August 2001.

25. TINA-C http://www.tinac.com.

26. Truyen E, Robben B, Kenens P, Matthijs F, Michiels S, Joosen W, Verbaeten P. Open implementation of a mobile communication system, Department of Computer Science, K.U. Leuven Technical Report www.cs.kuleuven. ac.be/~eddy/mp/smove.html.

5

Baseline Technology Review

This chapter reviews the main research results relevant to the programmable networks and their management.

5.1. Introduction

Programmable networks have been proposed as a solution for the fast, flexible, and dynamic deployment, customization and management of new network services. Programmable networks are networks that allow the functionality of some of their network elements to be programmed dynamically. These networks aim to enable the easy introduction of new network services by adding dynamic programmability to network devices such as routers, switches, and application servers. Dynamic programming refers to executable code that is injected into the network element in order to create the new functionality at run time. The basic idea is to enable third parties (end users, operators, and service providers) to inject application-specific services (in the form of code) into the network. Applications may utilize this network support in terms of optimized network resources and, as such, they become network aware. Programmable networks allow dynamic injection of code as a promising way of realizing application-specific service logic, or performing dynamic service provision on demand. As such, network programming provides unprecedented flexibility in telecommunications. However, viable architectures for programmable networks must be carefully engineered to achieve suitable trade-offs between flexibility, performance, security, and manageability.

Many concepts for programmable networks have been proposed recently. Depending on the literature, the definition of the programmable network varies. For example, Campbell *et al.* [21] define two schools of thought that have emerged to make networks programmable. The first one is spearheaded by the Opensig community that was established through a series of international workshops [38],

Fast and Efficient Context-Aware Services   Danny Raz, Arto Tapani Juhola,
Joan Serrat-Fernandez, Alex Galis   © 2006 John Wiley & Sons, Ltd

and the other, established by DARPA [26], constitutes a large number of diverse Active Network projects. Similar definitions can be found in the articles by Kalaiarul Dharmalingam and Martin Collier [44], where network programmability has been divided into the three categories: the Opensig initiative, Active Networks (AN), and Mobile Agents. An annual international conference – IWAN [52] – is dedicated to the research and development in the Active and Programmable networks technologies, systems, and services.

In the following sections, the AN concept is described in more detail.

ANs are packet-switched networks where packets can carry not only data but also code, or references to code, that will be executed at intermediate nodes as the packets propagate through the network. In the traditional network, each node performs only the processing necessary to forward packets towards their destination. In contrast, an AN is aware of the content of the packets that are flowing through it, and is capable of making customized modifications to the data within the packets. In other words, while traditional, 'passive' network function is store-and-forward, the AN is store-compute-and-forward. The AN's ability to modify traffic is based on active routers and switches that are able to perform customized computation on the messages streaming through them. In contrast, the routers in common use nowadays are only able to modify the packet headers, not the payload. Active network components are not restricted to operating in networks formed purely from active components, but can also coexist and interoperate with legacy routers, which transparently forward datagrams to the active parts of the network [36,79].

Two main approaches exist for the realization of active networks: Programmable Node and Encapsulation. In the first approach, programs are injected into the active node using a mechanism that is separate from normal data packet processing. With this approach, which uses existing network packet formats, the previously installed program is executed when data packets associated with it arrive at the node [36,79]. In contrast, using the Encapsulation approach a midget program is integrated into every packet, thus transforming existing data packets into capsules within the transmission frames. When a capsule arrives at an active node, an Execution Environment extracts, interprets, and executes the program. In this approach, the active node has built-in mechanisms to load the encapsulated code, an execution environment to execute the code, and permanent storage where programs can save and retrieve data.

Regardless of the approach chosen, the main idea is the same. The AN customizes the content of data packets, and the behavior of the network can be altered dynamically by injecting new programs into the network nodes. The ability to inject software into the nodes bestows such benefits as rapid application and protocol deployment, and rapid software updating and customization. These benefits are achieved by downloading user-specific programs and executing them when they are needed.

A number of international workshops have established the Opensig solutions while DARPA [26,27] has established a number of active networks projects. The two approaches have many common features and attempts have been made to combine them [36,88] in the form of programmable networks, whose features have become the main focus of standardization activities (i.e., IETF ForCES protocol working group [54]).

5.2. Open Signaling Approach

The term 'programmable networks' is used widely by the Opensig [65] community to characterize networks built on the principles of programmable interfaces as identified by the IEEE Project 1520 standards initiative. The IEEE P1520 group developed a reference model (RM) [17,47], which provides a general framework for mapping programming interfaces and operations of networks, over any given networking technology. The IEEE P1520 reference model defines the following four types of interfaces:

- CCM-interface (an NE interface). This connection control and management interface is a collection of protocols that enable the exchange of state and control information at a very low level between the network element and an external agent.
- L-interface (an NE interface). This defines an application program interface (API) that consists of methods for manipulating local network resources abstracted as objects. The abstraction isolates upper layers from hardware dependencies or other proprietary interfaces.
- U-interface (a network-wide interface). This provides an API that mainly deals with connection setup issues. The U-interface isolates the diversity of connection setup requests from the actual algorithms that implement them.
- V-interface (a network-wide interface). This provides a rich set of APIs to write highly customized software, often in the form of value-added services.

This reference model highlights network programmability from two viewpoints: service and resource specific, corresponding to the layers of the L abstraction model. This allows upper-level interfaces to program new network services using existing resource abstractions or to modify existing services using service-specific abstractions. The service-specific abstractions are built on generic resource abstraction. A third layer is introduced to enable network programmability, by means of composition, *via* a basic set of standard building-block abstractions [17,81], on which both the service-specific and resource layers are built.

5.3. IFTF ForCES Approach

- IETF Forwarding and Control Element Separation (ForCES) working group was formed recently with an objective similar to that of P1520, namely, 'to define a set of standard mechanisms for control and forwarding separation. The control and forwarding planes standardized separation mechanisms allows both planes to be developed and innovated in parallel while maintaining interoperability' [51,53]. The NE is a collection of components of two types: control elements (CE) and forwarding elements (FE) operating in the control and forwarding (transport) plane, respectively. CEs host control functionality such as signaling and routing protocols, whereas FEs perform operations on passing packets, such as metering, header processing, and scheduling. CEs and FEs may be interconnected in every possible combination (CE-CE, CE-FE, and FE-FE), thus forming arbitrary types of logical topology.
- CEs must discover the capabilities of the FEs before they can actually control them. This is achieved through an FE model [86]. One requirement of such a model mandates that the FE model should provide the means to describe existing, new or vendor-specific logical functions found in the FEs. A second key requirement for such a model is to describe the order in which these logical functions are applied in the FE [50].
- The ForCES FE model uses a building-block approach, which is very similar to the P1520 working group. It encapsulates logical functions by means of an entity called the 'FE block.' When this FE block is used outside the context of a logical function, it is viewed as equivalent to the base building blocks. When a user looks inside an FE block, then it is viewed as a resource building block. FE blocks are expected to form an extendable FE block library, which will be included in the ForCES standard and which would create the basis for designing and building complex NE behaviors. Such an FE model enables three levels of management and control for FEs: static FE, dynamic FE, and dynamic extensible FE control and configuration. The first level assumes that the structure of the FE is already known and fixed, the second level allows the CE to discover and configure the structure of the FE although selecting from a fixed FE block library, while the third level allows CEs to download additional functionality at run time, namely FE blocks into FEs.

5.4. DARPA Active Networks Approach

Active networks transform the store-and-forward network into the store-compute-and-forward network [21,23,72,90]. In a generic way the Active Network goals can be summarized as follows:

- To create an architecture that enables solutions for immediate and future needs to be conceived and implemented easily.
- To provide a quantifiable improvement in the number and applicability of services that can be provided in the network.
- To enable application-specified control of network resources.

The packets are active in the sense that they carry executable code together with their data payload. This code is dispatched and executed at designated (active) nodes that perform operations on the packet data, as well as changing the current state of the node to be found by the packets that follow.

Two approaches can be differentiated, based on whether programs and data are carried discretely, namely within separate packets (out-of-band) or in an integrated manner (in-band).

In the discrete case, the task of injecting code into the node and the task of processing packets are separated. The user or network operator first injects his or her customized code into the routers along a path. When the data packet arrives, its header is examined and the appropriate preinstalled code is loaded and executed to process the packet contents [31,83]. Separate code loading and execution mechanisms may be required for the control of code execution. This separation allows network operators to dynamically download code, which extends a node's capabilities and becomes available for execution by customers. At the other extreme lies the integrated approach, where code and data are carried by the same packet [50]. In this context, when a packet arrives at a node, code and data are separated, and the code is loaded to process the packet's data or change the state of the node. A hybrid approach has also been proposed [5].

Active networks have their own proposed reference architecture model [22] depicted in Figure 5.1.

Execution Environment: The Execution Environment is a virtual machine in which a user process is executed. EEs are the 'sterile' environments provided by the node to execute each user-defined service or function. A separate EE may be provided to each service invoked by a user. The EEs interact with the NodeOS through the Node Interface.

Node Operating Systems: NodeOS is needed to support multiple EEs simultaneously and to provide a base level of common functionality that can be used by the different EEs. The NodeOS is responsible for managing the resources at the node and providing an abstraction between the user processes and the underlying system resources. The system resources include CPU-cycles, memory, transmission bandwidth at the output physical links, etc. The NodeOS is responsible for resource allocation and has control mechanisms that separate the operation of the EEs from each other as shown in Figure 5.1. It is also responsible for isolation of the different user processes running at the node.

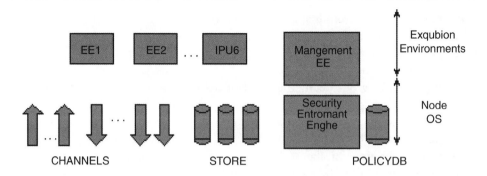

Figure 5.1   DARPA Active Node Architecture.

No user process can access resources and variables available to other processes. The NodeOS also ensures that no single process's use of resources hinders the performance of other user processes or the normal functioning of the node itself.

Links or Channels: These are the paths of communication between the NodeOS and the EEs, between the different EEs if allowed and between nodes. Packets or capsules traverse these links or channels.

The Active Network model describes an Active Network as a mixture of active and legacy nonactive nodes. The active nodes run the node operating system (NodeOS) – not necessarily the same at each node – while a number of execution environments may coexist at the same node. Finally, a number of Active Applications (AA) make use of services offered by the EEs. The AN reference architecture [23] is designed to simultaneously support a multiplicity of EEs at a node under the control of the NodeOS. Its major NodeOS functions are to isolate EEs from each other through resource allocation and control mechanisms, and to provide security mechanisms to protect EE and user data. Furthermore, only EEs of the same type are allowed to communicate with each other, while EEs of different types are kept completely isolated from each other. In addition, the NodeOS provides other basic facilities, such as caching or code distribution, that EEs may use to build higher abstractions to be presented to their AAs. All these capabilities are encapsulated by the node interface through which EEs interact with the NodeOS. This is the minimal fixed point at which interoperability is achieved [66]. In contrast, EEs implement a very broad definition of a network API ranging from programming languages to virtual machines such as the Spanner VM in smart packets and byte codes, to static APIs in the form of a simple list of fixed-size parameters [23]. To this end, an EE takes the form of a middleware toolkit for creating, composing, and deploying services.

5.5. Programmable Networks Components

The following sections describe in more detail the main technological components of programmable networks: Node Operating Systems, Execution Environments, and Management Systems.

5.5.1. Node OS: Node Operating Systems

The two distinct approaches for active networks are the pure capsule [49], which can be seen as an extreme in terms of how program code is injected into the network and the router plug-in [29] approach, which resembles a transition from upgradeable router architectures toward programmable, high-performance active network nodes.

In the capsules approach, every packet carries code that is executed at each node. Examples of capsule functionality are handling packet-routing requests or payload modifications to be carried out on a node. Capsules make use of a virtual machine that interprets the capsule's code to safely execute it on a node. The virtual machines restrict the address space, a particular capsule might access thus limiting the application of capsules, however this is done in order to ensure security.

A number of node operating systems (NodeOS) have been designed with the aim of controlling resources, as follows:

- *BOWMAN* [61] implements the active network interface specification in user space of Linux [56].
- *SCOUT* [62] is based on xkernel v2 and implements the path abstraction. It is a single address space research operating system without resource control mechanisms.
- *CROSSBOW* [31] follows the ideas of Scout and the path abstractions. Flows can be bound to plug-in chains. No resource control mechanisms are foreseen. Crossbow is strongly bound to a specific release of NetBSD [64].
- *EXOKERNEL* [27] multiplexes physical resources by providing a so-called library operating system that exports interfaces that are as close to the hardware as possible. As such it introduces as little as possible overhead. A clear disadvantage is code redundancy and the requirement to implement basic functionality found in legacy operating systems.
- *LARA* [76] provides an active node operating system that is implemented on Windows. It provides an execution environment, called a processing environment, into which active components are loaded. These components are interconnected to form a graph similar to the concepts of Scout.

- *MOAB* [89] is a research operating system based on the OSKit. It exports an interface as required for the Janos [45] operating system. Janos creates a Java virtual machine with resource control in mind.
- *NEMESIS* [55] is a research operating system for multimedia, low-latency communication. It introduced the fbuf structure to allow inter-process communication with zero-copy mechanisms.
- *PRONTO* [41] provides a framework for node programmability. It is based on Linux and runs in kernel space. It follows the plug-in approach by providing its own execution environment.
- *SILK* [8] is providing the path abstraction in the kernel space by extending the Linux kernel. Its architecture is similar to the Linux netfilter architecture for packet mangling.

Further work is needed in order to accommodate key characteristics that are missing from existing Node Operating Systems, such as:

- Management and control functionality that is not time critical and can be carried out in virtual machines without impinging on overall node performance.
- Simplification of the deployment and configuration of decoupled service components to eliminate extended time-consuming steps.
- Resource control for safely multiplexing physical resources on a node.
- High performance programmable network node architectures to support increased flexibility of application-specific code and to enable code processing at router link speed.
- Service composition and control. An interface to trigger service composition as well as its configuration should be available such that management and control components become reusable for different NodeOSs.
- Interfaces to specify resources in a very fine granular manner.
- Interfaces for intercomponent and inter-EE communication to ease the creation of decomposed services.

5.5.2. EE: Execution Environments

The architectural framework for Active Networks [22] defines a three-layer stack on each active node: NodeOS, Execution Environments (EE), and Active Applications (AA). The Execution Environments enable a programming model for designing Active Applications. They were also used for application-level networking [58,59,88]. In addition, a management EE [54] and management Active Applications are usually initiated at bootstrap, in order to offer management services at EE or AA level, respectively.

The following projects provide execution environments:

- *ANTS* [7] is a Java-based toolkit for constructing an Active Network and its applications. It is based on a capsule approach, in which code is associated with packets and run at selected IP routers that are extensible.
- *ASP* [19] is implementing the 'strong EE model' by offering a user-level operating system to the AAs *via* a Java-based programming environment. The underlying capabilities of NodeOS and Java are enhanced and so complex control plain functionality such as signaling and network management can be realized.
- *CANES* [24] supports high performance while also permitting dynamic modification of network behavior to support specific applications and/or provide new services. Its Execution Environments are built for composing services within the network [87] and run on the Bowman NodeOS [61].
- *FAIN* [36] introduced the virtual environment (VE) as a group of EEs. An EE type is characterized by the programming methodology and the programming environment that is created as a result of the methodology used. The EE type is free of any implementation details. In contrast, an EE instance represents the realization of the EE type in the form of a run-time environment by using specific implementation technology, for example programming language, and binding mechanisms to maintain its operation. The FAIN architecture also allows EEs to reside in any of the three operational planes, namely transport, control, and management, while they may interact and communicate with each other either across the planes or within a single plane. Within a VE, many types of EE with their instances may be combined to implement and/or instantiate a service [11,18]. A VE may coincide with an implementation (EE instance) that is based only on one technology, such as Java. Each node must have one privileged VE that is instantiated automatically when the node is booted up and serves as a controlled mechanism through which subsequent VEs may be created. This privileged VE should be owned by the network provider, who has access rights to instantiate the requested VE on behalf of a customer. This is achieved on the management plane by a VE manager (VEM). From this viewpoint, the creation of VEs becomes a kind of meta service.
- LANE [25] – Lighting Active Node Engine is an active network platform, comprising two main active components: Active Server (AS) and Active Router (AR); thus, the Execution Environment and the activity of the network are separated. Active Router communicates with Active Server through the Active Router API.
- *PLAN* [39] is a functional scripting language with limited capabilities, designed to execute on routers. The fundamental construct in the language is one of remote evaluation of delayed functional applications [40]. It has limited power to guarantee that all programs will terminate. It can also be used as a 'glue' layer

that allows access to higher level services. As such PLAN allows the flexibility of active networking, without sacrificing security.

- *SNAP* [73] is an active networking system where traditional packet headers are replaced with programs written in a special-purpose programming language. The SNAP language has been designed to be practical and with a focus on efficiency, flexibility, and safety. It offers significant resource usage safety, and achieves high levels of performance at the cost of flexibility and usability. A PLAN-to-SNAP compiler has also been developed [40].
- *TAMANOIR* [37,38] is based on Java language. Tamanoir active nodes provide persistent active nodes that are able to handle different applications and various data streams at the same time. Two main transport protocols: transmission control protocol (TCP) and user datagram protocol (UDP) are supported. The active network encapsulated protocol (ANEP) format is used to send data over active networks.

5.5.3. Programmable Management Services

This section provides an overview of the alternatives proposed for the management of programmable networks.

- *ANN* [28] – Active Network Node architecture makes use of active packets to deploy services. These packets feature a reference to a router plug-in, which contains the service logic. If not cached locally on a node, router plug-ins are fetched from a code server and installed on the node.
- *ANTS* [84] – Active Network Transport System uses active packets, which contain references to code groups, that is, the service logic. At the node level ANTS efficiently mimics an in-band deployment mechanism. As in the case of ANN, the choice between a centralized and a distributed approach is left to the service designer.
- *ANDROID* [4] – Active Network DistRibuted Open Infrastructure Development identifies an event- and policy-driven architecture for the management of Application Layer Active Networking (ALAN) [6,33] networks. It focused on the management of active servers, where programmability up to the application level is allowed.
- *ABONE MANAGEMENT* [1,2] is a DARPA-sponsored testbed for the active networks research program. It is configured as a number of virtual active networks. All nodes, which are administered locally, can be used by remote users to start up Execution Environments and Active Applications. Each user can run an instance of the active network management daemon – Anet. This daemon allows remote EE and AA developers to install, configure, and control EE instances in the testbed. Anetd performs two key management functions:

(i) demultiplexing of active network packets encapsulated using the Active Network Encapsulation Protocol to multiple EEs located on the same network node; (ii) deployment, configuration, and control of network software and EE prototypes.

- *ABLE* [3,48] – Active Bell Labs Engine addresses the management challenges of modern complex networks. It proposes an active engine that is attached to any IP router to form an active node, which executes programs that are oriented to the monitoring and control of the attached router. The active code is implemented in Java, and active packets are encapsulated in a standard ANEP header over UDP.
- *AVNMP* [10,34] – Active Virtual Network Management Prediction is a proactive management system achieved by modeling network devices within the network itself, and running that model ahead of real time. These predictions can concern either the network or an offered service. Simply, the algorithms compare the actual state of the network with previous predictions. If a previous prediction was incorrect, the configuration actions caused by this prediction are removed from the network. This correction is done through special kinds of messages called antimessages.
- *Brunner* [13–15] proposes a management system for multi-party active networks.
- *FAIN* [65] – Future Active IP Networks is using an extension of policy-based management for programmable networks. It also proposes component-based deployment and management of services in Active Networks [69].
- *Galis* [35] proposes a programmable management system for GRID services.
- *Fonseca* [32] defines a framework to allow capsule extensions of the policy-based management framework proposed by the IETF. Capsules represent user requirements and are used for service negotiation and network element configuration. Three types of capsule are permitted: capsules that request decisions from the policy execution point (PEP) to the policy decision point (PDP), capsules that notify decisions from the PDP to the PEP, and capsules that negotiate between ISPs.
- *Kato* [45] proposes a management framework designed to reduce management traffic by allowing network elements to make decisions. This is done by defining active packets, which might even contain policy parameters and code that are executed inside network elements. This allows network elements to make autonomous, intelligent decisions.
- *Haas* [42] proposes a new distributed service deployment over programmable networks.
- *Hicks* [40] uses a distributed, out-of-band approach to service deployment at the network level. It is targeted toward programmable networks, and, as a consequence, does not deal with code deployment.
- *Kanada* [43] proposes a method for the dynamic extension of a policy-based management system by means of policies in active networks. The method defines

two types of policy for realizing this extension: policy definition (PD) and policy extension (PE). A user can add a new type of policy into the policy server and can specify the corresponding methods for translating the new policy types into commands on different types of network nodes.

- *NESTOR* [63,85] is an architecture for NEtwork Self-managemenT and ORganization. The NESTOR system seeks to replace labor-intensive configuration management with the one that is automated and software intensive. It has been implemented in two complementary versions, and is now being applied to automate several configuration management scenarios of increasing complexity, with encouraging results.

- *SMART PACKETS* [77] focuses on applying active networks technology to network management and monitoring. The management applications developed are oriented to diagnostic reporting and fault detection. Smart packets are generated by management or monitoring applications and are encapsulated in ANEP. The ANEP daemon is responsible for receiving and forwarding smart packets correctly.

- *SERAPHIM* [74] enables the active code to dynamically install its own application-specific security functions. These code fragments, which are encapsulated inside active packets, have been named active capabilities (AC). An AC is able to carry not only the active code but also the security policies customized for a particular application, and even the code needed to make a policy decision.

- *SENCOMM* [70,75] – Smart Environment for Network Control, Monitoring and Management framework is an implementation of a network control, management, and monitoring environment using Active Networks, which reuses much of the smart packets system of [77].

- *PLANet/SwitchWare* [5] is an architecture that combines active packets with active extensions to define networked services. Active packets contain code that may call functions provided by active extensions. Active extensions may be dynamically loaded into the active node. At the network level, service deployment is implemented in a distributed, in-band way, using active packets similar to ANTS and ANN. Active packets contain code (in-band) that can be used like a glue to combine services offered by dynamically deployed active extensions (out-of-band). Similar to ANN and ANTS, the content of active packets may be modified by active extensions. Therefore, the choice between a centralized and distributed method of deployment is left to the service designer.

- *PBNM* – Policy-based network management is an emerging technology [68] for the management of telecommunications networks. PBNM can be adapted to manage Active Networks. Both the Internet Engineering Task Force [49,67] and the Distributed Management Task Force (DMTF) [30,57] are working on standards for policy-based network management.

- *VAN* [20] – Virtual active network management is a framework, which allows customers to access and manage a service in a provider's domain, and to outsource a service and its management to a service provider. Two types of EE exist in the management architecture: the management EE that operates on the management plane, and the service provider EE that operates on the data transfer and control planes. The tasks of the management EE are limited to node configuration and the management of virtual active networks in the active network provider's domain. In the VAN architecture, a service and the corresponding service management run in the same instantiation of a service provider EE.

References

1. ABone, http://www.isi.edu/abone/.
2. ABone – Introduction, http://www.isi.edu/abone/intro.html.
3. ABLE: The Active Bell Labs Engine, http://www.cs.bell-labs.com/who/ABLE/.
4. Active Network DistRibuted Open Infrastructure Development (ANDROID), http://www.cs.ucl. ac.uk/research/android/.
5. Alexander DS, *et al.* 'The SwitchWare Active Network Architecture.' *IEEE Network Special Issue on Active and Controllable Networks* 1998; **12**(3): 29–36. http://www.cis.upenn.edu/~switchware/papers/switchware.ps.
6. Application-Level Active Networks, http://dmir.it.uts.edu.au/projects/alan/.
7. AIN Release 1 Service Logic Program Framework Generic Requirements, Bell Communications Research Inc., FA-NWT-001132.
8. Bavier A, *et al.* SILK: Scout Paths in the Linux Kernel, Technical Report 2002-009, Uppsala University, February 2002.
9. Berson S, Braden B, Gradman E. The Network I/O Daemon–Netiod, October 11, 2001, Draft version. http://www.isi.edu/abone/DOCUMENTS/netiod.ps.
10. Bush SF, Kulkarni A. Active Networks and Active Network Management: A Proactive Management Framework. Kluwer Academic/Plenum Publishers, Norwell, MA: 2001.
11. Bhattacharjee S. 'Active Networks: Architectures, Composition, and Applications,' Ph. D. Thesis, Georgia Tech, July 1999.
12. Biswas J, *et al.* Proposal for IP L-interface Architecture, IEEE P1520.3, P1520/TS/IP013, 2000. http://www.ieee-pin.org/doc/draft_docs/IP/p1520tsip013.pdf.
13. Brunner M, *et al.* 'Management in Telecom Environments That Are Based on Active Networks.' *Journal of High Speed Networks*, March/April 2001.
14. Brunner M. 'Tutorial on active networks and its management.' *Journal Annals of Telecommunications*, 2002.
15. Brunner M, *et al.* 'Service management in multi-party active networks.' *IEEE Communications Magazine*, March 2000.
16. Biswas J, *et al.* 'The IEEE P1520 Standards Initiative for Programmable Network Interfaces.' *IEEE Communications*, Special Issue on Programmable Networks, 1998; **36** (10): http://www.ieee-pin.org/.
17. Bjorkman N, *et al.* 'The movement from monoliths to component based network elements.' Special Issue on Telecommunications Networking at the Start of the

21st Century, *IEEE Communications* 2001; **39**(1): http://www.msforum.org/techinfo/ IEEEcomMag200101_monoliths.pdf.

18. Braden B, *et al.* Introduction to the ASP Execution Environment (Release 1.5), November 30, 2001, http://www.isi.edu/active-signal/ARP/DOCUMENTS/ASP_EE.ps.

19. Braden B, Faber T, Handley M. 'From Protocol Stack to Protocol Heap–Role Based Architecture,' HotNets I, Princeton University, October 2002. http://www.cs.washington. edu/hotnets/papers/braden.pdf.

20. Brunner M, Plattner B, Stadler R, 'Service creation and management in active telecom environments.' *Communications of the ACM*, 2001.

21. Campbell AT, De Meer HG, Kounavis ME, Miki K, Vicente JB, VIllela D. 'A survey of programmable networks'. *ACM SIGCOMM Computer Communication Review*, 1999.

22. Calvert, KL (ed.). Architectural Framework for Active Networks, Draft version 1.0, July 27, 1999, http://protocols.netlab.uky.edu/~calvert/arch-latest.ps.

23. Calvert K, *et al.* 'Directions in Active Networks.' *IEEE Communications Magazine*, 1998, http://www.cc.gatech.edu/projects/CANEs/papers/Comm-Mag-98.pdf.

24. Composable Active Network Elements Project (CANES). http://www.cc.gatech.edu/ projects/ canes/.

25. CASPIAN project http://www.eurescom.de/public/projects/P900-series/P926/default. asp.

26. DARPA Active Network Program. http://www.darpa.mil/ato/programs/activenetworks/ actnet.htm.

27. 'DARPA Active Networks Conference and Exposition,' DANCE Proceedings – IEEE Computer Society Number PR01564, May 2002.

28. Decasper D, *et al.* 'A scalable, high performance active network node.' *IEEE Network*, 1999.

29. Decasper D, *et al.* 'Router Plug-ins: A Software Architecture for Next Generation Routers.' *Proceedings of ACM SIGCOMM '98*, Vancouver, Canada, September 1998.

30. Distributed Management Task Force, http://www.dmtf.org.

31. Decasper D, *et al.* 'A Scalable, High-performance active Network Node.' *IEEE Network*, 1999.

32. Fonseca M, Agoulmine N, Cherkaoui O. Active Networks as a Flexible Approach to Deploy QoS Policy-Based Management, http://citeseer.nj.nec.com/483138.html.

33. Fry M, Ghosh A. 'Application-level active networking.' *Computer Networks* 1999; **31**(7): 655–667.

34. Galtier V, *et al.* 'Prediction and Controlling Resource Usage in a Heterogeneous Active Network.' MILCOM 2001, October 2001.

35. Galis A, *et al.* 'Programmable Network Approach to Grid Management and Services.' International Conference on Computational Science 2003, LNCS 2658, June 2–4, 2003, pp. 1103–1113. www.science.uva.nl/events/ICCS2003/.

36. Galis A, Denazis S, Brou C, Klein C. (ed.) 'Programmable Networks for IP Service Deployment'. Artech House Books; www.artechhouse.com; ISBN 1-58053-745-6; pp. 450, June 04.

37. Gelas J, Lefevre L. 'TAMANOIR: A High-Performance Active Network Framework.' Active Middleware Services, Kluwer Academic Publishers, August 2000.

38. Gelas J-P, LefÉvre L. 'Toward the Design of an Active Grid.' Lecture Notes in Computer Science, Computational Science - ICCS 2002, Vol. 2230, April 2002, pp. 578–587.
39. Hicks M, *et al*. 'PLAN: A Packet Language for Active Networks.' *Proceedings of the Third ACM SIGPLAN International Conference on Functional Programming Languages, pp. 86–93, ACM, September 1998, http://www.cis.upenn.edu/~switchware/papers/plan.ps.*
40. Hicks M, Moore JT, Nettles S. 'Compiling PLAN to SNAP,' IWAN'01, September/October 2001. http://www.cis.upenn.edu/~jonm/papers/plan2snap.ps.
41. Hjalmtysson G. 'The Pronto Platform – A Flexible Toolkit for Programming Networks Using a Commodity Operating System,' OpenARCH 2000.
42. Haas R, Droz P, Stiller B. 'Distributed Service Deployment over Programmable Networks,' DSOM 2001, France, 2001.
43. Kanada Y. 'Dynamically Extensible Policy Server and Agent.' *Proceedings Policies for Distributed Systems and Networks*, 2002, pp. 236–239.
44. Dharmalingam K, Collier M. 'Netlets: A New Active Network Architecture.' First Joint IEI/IEE Symposium on Telecommunications.
45. Kato K, Shiba S. 'Designing Policy Networking System Using Active Networks.' Second International Working Conference on Active Networks (IWAN'2000), Tokyo, Japan, October 2000.
46. The Jasmin Project, http://www.ibr.cs.tu-bs.de/projects/jasmin/policy.html.
47. Standard for Application Programming Interfaces for ATM Networks, IEEE P1520.2, Draft 2.2, http://www.ieee-pin.org/pin-atm/intro.html.
48. Kornblum J, Raz D, Shavitt Y. 'The Active Process Interaction with Its Environment,' IWAN 2000, October 2000.
49. Internet Engineering Task Force, http://www.ietf.org.
50. Khosravi H, Anderson T. Requirements for Separation of IP Control and Forwarding, January 2003. http://www.ietf.org/Internet-drafts/draft-ietf-forces-requirements-08.txt.
51. IETF ForCES, draft-ietf-forces-framework-04.txt, December 2002. http://www.ietf.org/Internet-drafts/draft-ietf-forces-framework-04.txt.
52. IWAN (International Working Conference on Active and Programmable Networks) – www.iwan2005.net.
53. IETF ForCES. http://www.ietf.org/html.charters/forces-charter.html.
54. Jackson AW. *et al*. The SENCOMM Architecture, Technical Report, BBN Technologies, April 26, 2000. http://www.ir.bbn.com/projects/sencomm/doc/architecture.ps.
55. Leslie I, *et al*. The Design and Implementation of an Operating System to Support Distributed Multimedia Applications. http://www.cl.cam.ac.uk/Research/SRG/netos/old-projects/nemesis/documentation.html.
56. Linux. http://www.kernel.org.
57. Moore B. 'Policy Core Information Model (PCIM) Extensions,' RCF3460, January 2003.
58. Marshall IW, Roadknight, CM. 'Provision of quality of service for active services.' *Computer Networks* 2001; **36**(1): 75–87.
59. Marshall I, *et al*. 'Application-level programmable internet work environment.' *BT Technology Journal* 1999; **17**(2): 82–95.

60. Mathieu B, *et al.* 'Deployment of Services into Active Networks.' *Proceedings of WTC-ISS* 2002, Paris, France, September 2002.
61. Merugu S, *et al.* 'Bowman: A Node OS for Active Networks.' *Proeedings of IEEE Infocom* 2000, Tel Aviv, Israel, March 2000. http://www.cc.gatech.edu/projects/CANEs/papers/bowman.pdf.
62. Montz A, *et al.* 'Scout: A Communications-Oriented Operating System.' IEEE HotOS Workshop, May 1995.
63. Nestor Project: http://www1.cs.columbia.edu/dcc/nestor.
64. NetBSD. http://www.netbsd.org.
65. Open Signaling Working Group. http://www.comet.columbia.edu/opensig/.
66. Peterson L (ed.). Node OS Interface Specification, AN Node OS Working Group, November 30, 2001. http://www.cs.princeton.edu/nsg/papers/nodeos-02.ps.
67. Resource Allocation Protocol IETF's WG, http://www.ietf.org/html.charters/rap-charter.html.
68. Sloman M, Lupu E. 'Policy Specification for Programmable Networks.' International Working Conference on Active Networks (IWAN'99), Berlin, Germany, June-July 1999.
69. Solarski M, Bossardt M, Becker T. 'Component Based Deployment and Management of Services in Active Networks.' Proceedings of Fourth Annual International Working Conference on Active Networks (IWAN 2002), ZÏrich, Switzerland, also in Lecture Notes in Computer Science 2546, Springer Verlag, December 2002.
70. Smart Environment for Network Control, Monitoring and Management, http://www.ir.bbn.com/projects/sencomm/sencomm-index.html.
71. Stevenson DW. 'Network Management: What It Is and What It Isn't,' white paper, April 1995.
72. Smith JM, *et al.* 'Activating networks: A Progress Report.' *IEEE Computer* 1999; **32**(4): 32–41. http://www.cs.princeton.edu/nsg/papers/an.ps.
73. SNAP: Safe and Nimble Active Packets, http://www.cis.upenn.edu/~dsl/SNAP/.
74. Seraphim Project homepage, Seraphim: Building Dynamic Interoperable Security Architecture for Active Networks, http://choices.cs.uiuc.edu/Security/seraphim/.
75. Sencomm Project: http://www.ir.bbn.com/projects/sencomm/sencomm-index.html.
76. Schmid S, *et al.* 'Flexible, Dynamic, and Scalable Service Composition for Active Routers.' Proceedings Fourth Annual International Working Conference on Active Networks (IWAN 2002), ZÏrich, Switzerland, Lecture Notes in Computer Science 2546, Springer Verlag, December 2002.
77. Schwartz B, *et al.* 'Smart Packets for Active Networks,' OpenArch '99, March 1999.
78. Tullmann P, Hibler M, Lepreau J. 'Janos: A Java-oriented OS for active networks.,' *IEEE Journal on Selected Areas of Communication* 2001; **19**(3): 501–510.
79. Tennenhouse D, Jonathan S. 'A survey of Active Network Research'. *IEEE Communications Magazine*, 1997.
80. Van der Merwe JE, *et al.* 'The Tempest – A Practical Framework for Network Programmability.' *IEEE Network*, Vol. 12, No. 3, May/June 1998, pp. 20-28. http://www.research.att.com/~kobus/docs/tempest_small.ps.
81. Vicente J, *et al.* L-interface Building Block APIs, IEEE P1520.3, P1520.3TSIP016, 2001. http://www.ieee-pin.org/doc/draft_docs/IP/P1520_3_TSIP-016.doc.

82. Vicente J, *et al.* 'Programming Internet Quality of Service,' 3rd IFIP/GI International Conference of Trends toward a Universal Service Market, Munich, Germany, September 12–14, 2000. http://comet.ctr.columbia.edu/~campbell/papers/usm00.pdf.

83. Wetherall D, *et al.* 'ANTS: A Toolkit for Building and Dynamically Deploying Network Protocols.' Proceedings of IEEE OPENARCH '98, April 1998.

84. Wetherall D, *et al.* 'ANTS: A Toolkit for Building and Dynamically Deploying Network Protocols.' Proceedings of IEEE OPENARCH '98, April 1998.

85. Yemini Y, Konstantinou A, Florissi D. 'NESTOR: An architecture for NEtwork Self-management and Organization.' *IEEE Journal on Selected Areas in Communications* 2000; **18**(5): 758–766.

86. Yang L, *et al.* ForCES Forwarding Element Functional Model, March 2003.

87. Zegura E (ed.). Composable Services for Active Networks, AN Composable Services Working Group, September 1998. http://www.cc.gatech.edu/projects/CANEs/papers/cs-draft0-3.ps.gz.

88. Karnouskos S, Guo H, Becker T. 'Trade-Off or Invention: Experimental Integration of Active Networking and Programmable Networks.' Special Issue on Programmable Switches and Routers. *IEEE Journal of Communications and Networks* 2001; **3**(1): 19–27.

89. Moab. http://www.cs.utah.edu/flux/janos/moab.html.

90. Wetherall D, Tennenhouse D. 'The ACTIVE IP Options.' Proceedings of the 7th ACM SIGOPS European Workshop, September 1996.

6

CAS Creation and Management – System Architecture and Design Considerations

In the preceding chapters the various aspects of context-aware services were considered, including their nature, operation, necessary functional architecture, and technological prerequisites. This chapter outlines the system components supporting the required functionality, discusses their design, and presents the relations between them. The resultant system architecture covers the aspects of flexible definition, dynamic customization, automated provisioning, and maintenance of services making use of context information.

6.1. Introduction

From the technical point of view the CONTEXT solution spans into three domains, namely the Service Layer (SL) domain, the Active Applications Layer (AAL) domain, and the IP domain.

The service layer domain deals with the modeling of the information expressing the context of services, the modeling of the services themselves, and the required framework for the service creation and management. The AAL-based solution, with appropriate APIs to control the IP domain, allows the actual delivery and management of the context-aware services. The control of the IP domain typically includes configuring routers to intercept-specific packet types, or to alter routing, traffic shaping, and QoS settings. The division of functionality into AAL and IP domains is important; because the lion's share of network-related services beyond packet delivery can be deployed without affecting the behavior, that is code, of IP-packet processing in routers. Being separated from the IP layer promotes

Fast and Efficient Context-Aware Services   Danny Raz, Arto Tapani Juhola,
Joan Serrat-Fernandez, Alex Galis   © 2006 John Wiley & Sons, Ltd

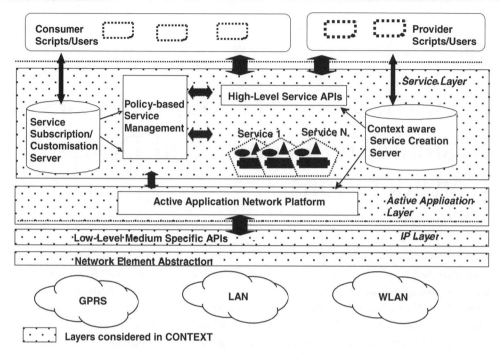

Figure 6.1 Overall CONTEXT System Architecture.

robustness—any mistakes at the IP-layer are likely to cause serious damage—and efficiency since the IP-layer remains simpler and faster. This approach is also known as 'Application Layer Active Networking,' ALAN [1], and, although having different starting points, the notion of ALAN is fairly close to the idea of Mobile Agents [5,6].

Figure 6.1 shows the domains of CONTEXT system and involved entities: Users above and network technologies below. Under AAL we find medium-specific APIs and network element abstractions. The issue here is to unify the differences between media and network elements, so that the AAL code can be isolated from the IP layer implementation details.

The CONTEXT solution has been coined as 'ContextWare' Programmable Middleware [3] due to its position between service customers, service providers, and the underlying transport network.

6.2. Service Layer Overview

The Service Layer (SL) domain is about the modeling of context-aware services and the context information involved, and the creation and management of such

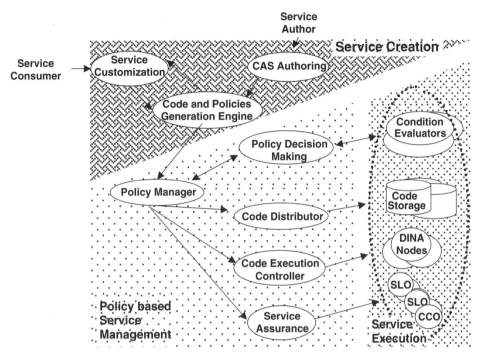

Figure 6.2 Service Layer Architecture (SLO = Service Level Object, CCO = Context Computational Object).

context-aware services. Figure 6.2 presents the functional division of the service layer domain, demarcated with polygons, and superimposed on them is depicted the associated system architecture. Note that many functional components showed in this figure were already presented in Figure 3.3 in order to introduce the service life cycle of CASs. In addition to the previously described components, we have now others dictated by the technology that has been used for the service layer implementation architecture, namely the Policy-based Management approach for Service Management and the DINA active platform for the Service Execution. Therefore, we are going only to introduce the functionality of these new components and we refer the reader to Chapter 3 for the description of all the other components.

6.2.1. Policy Management Components

The Policy Framework is of generic nature and its purpose is to cater for Policy-Based Service Management functionality. The framework involves the following components:

- *Policy Manager* (PM), for receiving and managing policies, that is seeing that the prerequisites of the enforcement of the policies are fulfilled, that the policies are followed in all locations where they should, and that only up-to-date policies are present in the system.
- *Policy Decision-Making component* (DMC), responsible for evaluating policies, in other words assessing the system state changes against the policy rules and initiating suitable reactions if required.

Various *Action Consumer components* (AC), responsible for undertaking specific actions under the instructions of the Policy Manager—which in turn bases its instructions on the decisions of the Policy decision-making component. For the CONTEXT system the following functions were realized by the respective action consumers:

- The Code Distributor AC, intended to implement the Code Distributor functionality.
- The Code Execution Controller AC, intended to implement the Code Execution Controller functionality.
- The Service Assurance AC, intended to implement the Service Assurance functionality.

These actions may be proactive or reactive. Proactive measures aim to configure the system in order to better serve future requirements. One example of proactive measures is the production of policies altering code execution tactics, so as to direct future service activation requests to less utilized nodes of the execution environment or to deny specific service activations. Reactive measures aim to rectify the current operation of the system in order to improve the performance of currently executing service code. One example of reactive measures is the real-time configuration of individual operating service code or even terminating specific service code in order for the rest to achieve the required performance.

6.2.2. Service Execution Components

The Execution Environment of the CAS code is distributed and based on the DINA platform, in essence the active nodes of an IP network.

As noted earlier, the code will not be installed straight away into all active nodes. Very likely it will be stored into *Code Storage components* to wait to be fetched as need arises. This is required for the maintenance of the code base and for the sake of user and code mobility. The storage point may be a dedicated server or a DINA node. Decision on the storage location belongs to the policies defined for each service.

The *Condition Evaluator components* (CE) are responsible for providing the Policy decision-making component with the necessary information for evaluating policies. Each time a new policy is received by the Policy manager the relevant Condition Evaluators that provide info for evaluating this policy must be configured/ activated.

The Condition Evaluators employed by the policy framework exist in the form of active code executed in DINA nodes, receiving information from node interfaces made available to them. Specifically, the interfaces are offered by the nodes' internal DINA Brokers, gathering, mediating, and standardizing the exchanges of information for the benefit of active code. The pieces of active code, in this case Condition Evaluators, also report events back to the Policy decision-making component via these brokers, as configured per service policy.

6.2.3. Interfaces Between Service Layer Components

Taking into consideration the necessary functional components as presented in Figure 6.2, we adopted the decision to make a physical architecture consisting in a one-to-one relationship between functional and physical components; that is, each function is implemented by means of one component. What follows here is an account of the relations of the service layer domain components so far introduced, including the relevant information they exchange, that is interface descriptions. The interface descriptions are ordered to mirror the service life cycle and therefore they consist of the implementation of the reference points considered in the Service Layer Reference Points of Figure 3.4. An overview of the listed interfaces is shown in Figure 6.3.

6.2.3.1. CAS Authoring

(Input) System Capabilities: The available system capabilities for constructing Context-aware Services, based on the CAS modeling approach.

(Input) CAS Author: The input from the person responsible for defining services.

(Output) Service Definition Document: An implementation technology-independent document describing all details required for a service's creation and provisioning.

6.2.3.2. Service Customization

(Input) Consumer request: A consumer's request for a service, containing all the necessary service customizations representing the consumer's individual needs.

(Input) Customizations Interface Configurations: A document specifying the customizations interface exposed to the consumers for the specific service—this document is based on the capabilities of the front-end technology implementing the

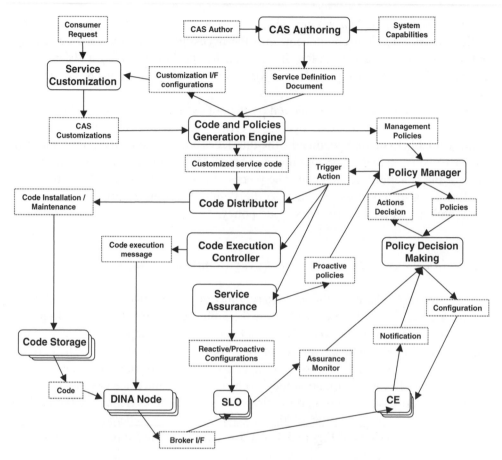

Figure 6.3   Service Layer Components' Interfaces.

interface, for example, dynamic web pages and a set of policies realizing the back-end logic of the Customization server.

(Output) CAS Customizations: A document containing the consumer's specific customizations for the service.

6.2.3.3. Code and Policies Generation Engine

(Input) Service Definition Document: Received from the CAS Authoring component.

(Input) CAS Customizations: Received from the Service Customization component.

(Output) Customized Service code: The actual code that implements the logic of the service fulfilling the demands of a service subscription. The implementation technology of the service code is compatible with the technologies supported by the execution environment.

(Output) Management Policies: Management policies are policies resulted from the definition epoch, pertinent to the service, and also policies resulted from the customization epoch, pertinent to specific customizations of the service. These policies will configure the policy-based management system for sufficiently managing the service.

6.2.3.4. Policy Manager

(Input) Management Policies: Received from Code and Policies Generation Engine.
  (Input) Action Decisions: As decided by the Policy decision-making component.
  (Input) Proactive Policies: Received from the Assurance component.
  (Output) Policies: To be evaluated by the Policy decision-making component.
  (Output) Action Triggers: Corresponding to Action decisions aimed at the relevant action consumers.

6.2.3.5. Policy-Decision Making

(Input) Policies: Received from Policy manager.
  (Input) Notifications: Received from the condition evaluator corresponding to the necessary information for the policies under evaluation.
  (Input) Assurance Monitor: Received from the SLO corresponding to the necessary information for assurance policies under evaluation.
  (Output) Action Decisions: Based on the information received by the relevant condition evaluators and on the policies under evaluation, action may be decided. These decisions are forwarded to the policy manager for their enforcement.
  (Output) Configuration: Configuration info for the corresponding condition evaluators, so as to support the information gathering for the policies under evaluation.

6.2.3.6. Code Distributor

(Input) Action Trigger: Received from Policy manager.
  (Input) Customized Service code: Received from the Code and Policies Generation Engine.
  (Output) Code Installation and Maintenance: Is comprised by actions that perform the optimal installation of the customized service within the storage points of the execution environment and actions that ensure the reliable maintenance of the installed code.

6.2.3.7. Code Storage

(Input) Code Installation/Maintenance: File managing instructions from the code distributor.
  (Output) Code: Code download capabilities using appropriate code URL.

6.2.3.8. Code Execution Controller

(Input) Action Trigger: Received from Policy manager.

(Output) Code Execution Message: Is a message dictating the execution of a specific customized service code, at specific node(s) of the execution environment with specific run time arguments.

6.2.3.9. Service Assurance

(Input) Action Trigger: Received from Policy manager.

(Output) Reactive/Proactive Configurations: Configurations of the operating code aiming at optimizing its performance and/or the performance of the overall system.

(Output) Proactive Policies: Policies that will alter the configuration of service management. These policies will affect code execution and code distribution mechanisms.

6.2.3.10. Condition Evaluator

(Input) Configuration: Configuration instructions received from the Policy decision-making component.

(Input) Broker I/F: Information retrieved from Broker interfaces as offered by DINA.

(Output) Notifications: To the Policy decision-making component for evaluating corresponding policies.

6.2.3.11. DINA node

(Input) Code: Service code downloaded from storage points given its URL.

(Input) Code Execution message: Message requesting the execution of code as an active application, containing the URL of the code and its run-time arguments.

(Output) Broker I/F: Exports the capabilities of the underlying network infrastructure through the Broker APIs.

6.2.3.12. SLO

(Input) Broker I/F: Is able to utilize the capabilities of the underlying network infrastructure through the Broker APIs.

(Input) Reactive/Proactive Configurations: Received from the Service assurance component.

(Output) Monitoring Data: Self-monitoring to the Policy decision-making component.

6.3. Service Layer Implementation Considerations

According to what was mentioned in Chapter 3 and in the above sections the CAS Creation component, the Service Customization, and the Code and Policies Generation Engine are implemented as stand-alone applications, integrating all necessary tools that facilitate the creation of services and offering to the administrator a friendly graphical interactive interface.

As for the specific technologies, the modeling of CAS uses XML and the XML schema specifications. For implementing the GUIs Java technology is used, including swing libraries and XML parsers. The XML parsers employed by the GUI and other components discussed below are based on available XML parsing libraries. The repositories containing the CAS info (policies, module documents, etc.) are implemented using database technology. For handling the flow of information between the components associated with Authoring, Customization, Policy-based management, and Code and Policy Generation Engine SOAP is used. The Code Generator component is bound to Java technology in order to produce DINA-compatible code, and the produced code is placed into an Apache server to make it available to the code distributor component.

The policy-based management paradigm that is the foundation of the service management implementation has been touched earlier. Now it is time to shed some light on the rationale and impact of using this approach.

6.3.1. Why Policies?

A policy is an administrator-specified directive that manages and provides guidelines for how the different network and service elements should behave when certain conditions are met. However, the answer to the question, 'why policies should be used?' is a justification itself and explains why the IST-Context consortium decided to use policies as the main engine to create, deploy, and manage context-aware services.

The main benefits from using policies are improved scalability and flexibility for the management system. Scalability is improved by uniformly applying the same policy to large sets of devices and objects, while flexibility is achieved by separating the policy from the implementation of the managed system. Policies can be changed dynamically, thus changing the behavior and strategy of a system, without modifying its implementation or interrupting its operation. Policy-based management is largely supported by standards organizations as IETF and DMTF, and most network equipment networks.

Another benefit from using policies for management is their simplicity. This is achieved by means of two basic techniques: Centralized Configuration, you do not have to configure each element individually; and Simplified Abstraction, which means that you do not have to configure exactly each device, but only establish the policy that you want the overall system to follow and the system will translate this policy for you and will enforce it in the correct component.

As will be explained in this document, the policies are predefined in XML; we have chosen XML because it is a common language, standardized, and versatile. When a user subscribes to an available CAS, then policies are personalized taking these predefined policies as a template and taking into account the user's preferences and context.

It is proposed the use of XML as the language to express policies and use the proposed information model and architecture to manage the services based on these policies. The main advantage of using XML as policy language is its flexibility to define and exchange policies written in this format.

6.3.2. Objectives of the Policy-Based Service Management System

The main objective of using policies for service management is the same as for managing networks with policies: We want to automate management and do it as high level as possible. The philosophy for managing a resource, a network or a service with a policy-based managed approach is that IF something happens THEN the management system takes an action.

Policies can be tailored to different users. The main idea is to use generic policies that can be customized, following user subscription; the parameters of the conditions and actions in the policies are different for each user, reflecting its personal characteristics and its desired context information.

Within the CONTEXT system, the application of policies encompasses the expression and subsequent creation of portions of the logic of context-aware services. This is undertaken by the activities of the CAS Creation subsystem.

In addition, policies are used in managing various aspects of the created context-aware services. An important aspect of policy-based service management is the deployment of services throughout the Active Network. For instance, when a context-aware service is going to be deployed, the code storage points must be decided, based on the policies customized according to the values introduced by the user and network context or environment. Furthermore, context-aware service invocation and execution is also controlled by policies, for determining in a flexible way when, where, and how customized service code will be invoked and executed. Finally, the maintenance of the code realizing the logic of context-aware services and the assurance of context-aware service operation is also subject to related management policies.

The above concepts guided our decision to use predefined policies expressed in XML, which can later be personalized by user subscriptions to the context-aware services. The solution for the Policy-Based Service Management of the IST-Context

Project must rely on a robust and flexible architecture to accommodate the service management systems and new types of services. The architecture should exhibit the following logical attributes:

- Open: It is built on standard interfaces between architectural components.
- Flexible: It supports the incorporation of any new context services that complies with the specified interfaces, and allows the modification of the system behavior by means of user-defined policies.
- Modular: Its architecture is based on components, standardized, and oriented to context services.
- Scalable: It separates in different building blocks of technology-specific and technology-independent functionality, and minimizes data duplication.
- Distributed: It comprises a component-based architecture defined to run on top of a standards-based object request broker guarantees the transparency of the system to the location of components.

6.4. Context Policy-Based Service Management System

After introducing the framework that is going to be the basis of the design, we can now propose the Service Management Layer Architecture. We will apply the concepts of the generic policy-based management system (PBMS) in order to specify and implement the required functionality for the Service Management Layer. To do this we will adapt the mechanism described here to the CONTEXT problem, proposing new specific components and matching the functionalities identified with the philosophy of a generic policy-based management system.

When designing the Policy-Based Service Management Layer for CONTEXT project, the different functional blocks identified earlier were taken as a reference. As stated in this document, the result of the Service Creation phase is the Service Code (Java) and the Management Policies (XML) that are going to manage the provisioning and the maintenance of this service. As the work focuses on defining the Policy-Based Service Management Layer, the Service Creation Phase is not affected, as the Service Management layer only concerns the interaction between the two layers or phases.

Regarding Service Management Layer, there are four main functional blocks identified. They are:

- Code Distribution and maintenance
- Code Execution
- Service Invocation
- Service Assurance

As stated above, these functional blocks must be policy driven. So the next step is to design a policy management system that handles, manages, and applies the service management policies that will rule the behavior of the system, and particularly the efficient delivery of the functionalities identified. This step obviously includes the proposal of the classes or types of policies that will be managed by the system and its components.

The components that are technology and policy specific are the Action Consumers and the Condition Evaluators. These components are responsible for interpreting the particular policy semantics, monitoring its conditions (CE), and applying its actions (AC). Thus it is reasonable that in order to provide the functionalities we should use specific Action Consumers and Condition Evaluators proposed for this objective.

So, the idea is to propose the necessary Action Consumers and Condition Evaluators in order to cater for every identified functionality. Also, the policies proposed will be tightly coupled with these functionalities. Based on these premises, our initial implementation architecture is shown in Figure 6.4.

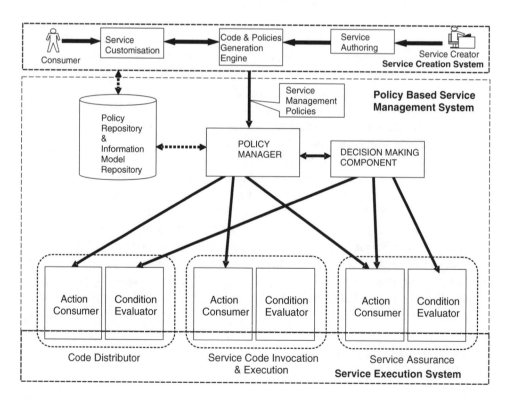

Figure 6.4   First Approximation of the Generic Policy-Based Management System to the CONTEXT Service Layer Functional Requirements.

The AC and CE, which are grouped according to the functional components they are related to, are assigned to system components as follows:

- Code Distributor system component: Code Distributor Action Consumer and Code Distributor Condition Evaluator
- Service code Invocation & Execution system component: Code Execution Controller Action Consumer and Service Invocation Condition Evaluator
- Service Assurance system component: Condition Evaluator and Service Assurance Action Consumer.

The Condition Evaluators proposed may not be a unique component but a set of different components specialized for particular tasks installed in different points of network. For example, for service assurance there could be a number of different Condition Evaluators instances, each specialized for monitoring different performance or quality parameters, or installed in different nodes. Another example could be that for service invocation there could be different Condition Evaluators, each specialized for monitoring or listening to different kinds of invocation signals, or events and installed in different points.

Finally, the proposed architecture for the Service Management layer can be seen in Figure 6.5.

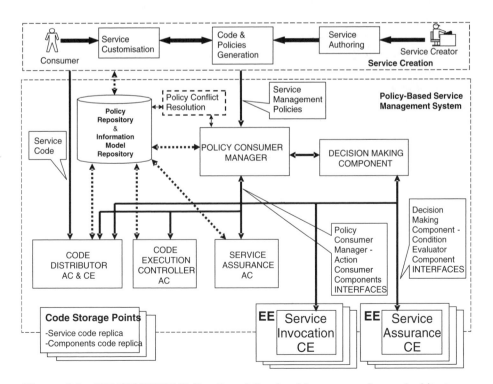

Figure 6.5   IST-CONTEXT Policy-Based Service Management Layer Architecture.

This architecture is the result of the adaptation of the policy-based management system described earlier to the necessities and functional requirements of the CONTEXT platform.

6.4.1. On System Components

The specific CONTEXT components introduced in the architecture playing the role of Action Consumers (AC) or Condition Evaluators (CE), or giving support to CAS management functional blocks are briefly described in the following sections.

6.4.1.1. Code Distributor Action Consumer

The Code Distribution Action Consumer (CDAC) distributes the active code ultimately destined for active nodes. Due to the dynamic nature of the system the exact node or nodes might not be known until the need arises. In such cases intermediate storage locations must be selected and used. Also, to simplify code life-time management, the count of code instances in circulation must be kept at minimum. CDAC enforces the actions related to the code distribution policies that have been sent by PM. The result of the actions will be the distribution of the service code following specific configuration parameters or Code Storage Point selection criteria included in the policy action. CDAC must be aware of network capabilities information in order to perform its actions, so it is able to communicate with the Information Repository to obtain the required information.

The Code Distributor Action Consumer will run outside the active nodes. This is because the wide scope of its actions; the count of the Storage Points might be large. The Code Distributor is a Java class; the API offered by it is used by Policy Manager to invoke the different code distribution-related actions contained in the Code Distribution Policies.

Code Distributor Action Consumer API

The main methods are the following:

- 'DistributeCode' method provides the mechanism to distribute a piece of active code, that is a (Service Layer Object)(SLO), (Context Computational Object) (CCO) or (Service Invocation Condition Evaluator) (SICE) to a set of code repositories or code execution points (DINA nodes) depending on the 'level' argument. The level refers to administrative domains, for example level 1 as close to the DINA nodes as possible and level 3 not very close at all. This method also adds the entry for the new code into the code_storage table.
- 'GetOptimalURLOfCode' method provides the mechanism to retrieve a specific URL of where the code resides by looking it up in the code_storage table.

- 'RemoveCode' method provides the mechanism to remove a piece of code from a specific code repository or DINA node along with its corresponding entry in the code_storage table.

```
public class CodeDistributorInterface {

    public CodeDistributorInterface();

    public boolean DistributeCode(String CodeId, URL[]          URLList,

int NoCopies, int Level, String[]        PotentialExecPoints);

    public URL[] GetOptimalURLOfCode(String CodeId,

            InetAddress DINANodeIPAddr);

    public boolean RemoveCode(String CodeId);
}
```

6.4.1.2. Code Execution Controller Action Consumer

The Code Execution Controller Action Consumer (CECAC) is the specific component for the control of the execution of the Service Code. It means that the actions that will be enforced by these consumers are related to when and where to execute the service code. For example, after receiving an invocation signal the actions that start the service execution in the appropriate execution points are executed by this Action Consumer.

The actions of the Code Execution Action Consumer affects many nodes (Execution Points), so it was deemed sensible not to locate it in them. The Code Execution Controller is a Java class. The Code Execution Controller offers an API to be used by the Policy Manager to invoke the different code execution-related actions contained in the Code Execution Controller Policies.

Code Execution Controller API

'executeCode' triggers an execution of a service that is identified by the 'codeId' and passes the arguments 'arg' to this service. The code can be executed on a specific node ('node') or on a set of nodes that are derived from a wildcard ('nodeWildcard'). The methods return '0' on success or '−1' in case of a failure.

'sendMessageToService' sends the message 'msg' to a running service that is identified by 'codeId'. The message can be sent to a service that runs on a specific node ('node') or on a set of nodes that are derived from a wildcard ('nodeWild-card'). The methods return '0' on success or '−1' in case of a failure.

```
public class codeExecutionController
{
    public codeExecutionController();
```

```
public int executeCode(String codeId, InetAddress
      node, String [] arg);
public int executeCode(String codeId, String
      nodeWildcard, String [] arg);

public int sendMessageToService(String codeId,
      InetAddress node, String msg);
public int sendMessageToService(String codeId,
      String nodeWildcard, String msg);
}
```

6.4.1.3. Service Assurance Action Consumer

Once the service code has started its execution, it is time to proceed to its Assurance. The Service Assurance Action Consumer (SAAC) executes the actions related to the Service Assurance, which is invoked when certain performance or assurance conditions are met, as determined by Service Assurance Condition Evaluators.

The Service Assurance Action Consumer runs outside the active nodes; its actions are not reduced to the scope of any particular active node. The service assurance functionality is treated in more detail later on in this chapter.

6.4.1.4. AC Data API

The different AC must share diverse run-time information which must be stored somewhere in the system. To provide a common interface, as well as information integrity, the AC DATA API is provided.

Its goal is to store, maintain, and delete information based on explicit requests of the different ACs. Actually, the AC DATA API comprises four APIs, to manage four different data tables. The tables are described in the following paragraphs, followed by the details of the APIs. The tables are:

CODE_PROCESS Table: To register information of the current execution processes over the DINA network. It contains the following information:

- CodeID
- DINASessionID
- DINASeqNumber
- DINANodeID   (multiple)
- DINANodeIPAddress   (multiple)

The name of the accompanying API is `codeProcessHandler`.

CODE_STORAGE Table: To keep track of the different locations where different copies of code are stored. It contains the following information:

- CodeID
- Copy N°
- URLList: list of URLs which contain the code

The name of the accompanying API is `codeStorageHandler`.
DINA_NODES Table: To store information relevant to the DINA nodes in the network. It contains the following information:

- DinaNodeID
- DinaNodeIPAddress
- Location: Physical location of the DINA node
- HasList: Contains information of the components/devices the DINA node has (i.e., SIPBroker, WLANBroker, WLANmodel=Cisco74, etc.)

The name of the accompanying API is `dinaNodeHandler`.
STORAGE_POINTS Table: To store information relevant to the DINA nodes in the network. It contains the following information:

- storagePointID
- storagePointIPAddress
- Protocol
- Port
- BackupLevel
- ListofDinaNodes

The name of the accompanying API is storagePointHandler.
The APIs to create, alter, access, and delete information from the above tables are specified as follows:

CodeProcessHandler API:

```
public interface codeProcessHandler {

  // New process
  public void addNewProcess(String codeId, int sessId,
        int seqNumber, InetAddress dinaNodeIp);

// Recover Information
public InetAddress [] getDinaNodesIPs (String
                    codeId);
public int getSessionId(String codeId);
public int getSequenceN(String codeId);
```

// Delete
public void removeProcess (String codeId);
public void removeProcessFromNode (String codeId,
 InetAddress dinaNodeIP);
}

Comments about codeProcessHandler API:

- removeProcess () removes the process specified by the codeID from all DINA nodes
- removeProcessFromNode () removes the process only from the specified node (given by the IP or the dinaNodeID)

CodeStorageHandler API:

public interface codeStorageHandler {

// New code
public void addCode(String CodeId, int copyNumber,
                                        String[] URLList);

// Recover info
public String[] getStorageURLs(String CodeId);
public int[] getCopyNumbers(String CodeId);

// Delete
public void removeCode(String CodeId);
}

Comments about codeStorageHandler API:

- addCode() usage example: If CD AC wants to distribute a certain code of a service named "SRV1" which has two classes "main.class" and "aux.class", CD should use this method in a similar way as below:

addCode ("SRV1", 1, {"ftpA://code/srv1/main.class",
"ftpA://code/srv1/aux.class"});
   addCode ("SRV1", 2, {"ftpB://opt/srv1/main.class",
"ftpB://opt/srv1/aux.class"});
   addCode ("SRV1", 3,
                    "ftpC://context/srv1/main.class",
                          "ftpC://context/srv1/aux.class"});

NOTE: `copyNumber` can be seen as a copy identifier, and it should be different for all copies

dinaNodeHandler API:

public interface dinaNodeHandler {

// New node
public void addDinaNode(String nodeID, InetAddress   nodeIP,
String location, String [] HasList);

// Recover info
public InetAddress[] getDinaNodes(String Location, String []
FilteringWildcard);
   public InetAddress getDinaNodeIP(String dinaNodeID);
   public String getDinaNodeID(InetAddress dinaNodeIP);

  // Delete
  public void removeDinaNode(String NodeID);
  public void removeDinaNode(InetAddress NodeIP);
}

Comments about dinaNodeHandler API:

- `HasList` passed as parameter in the `addDinaNode ()` method is similar to the following:

  HasList1={WLANbroker=yes, WLANmodel=Cisco62, SIPBroker=yes}

- The `FilteringWildCard` parameter in the `getDinaNodes ()` method is a string array. Each element of the array has the following format:

element=attribute

  e.g.:

  NodeID1 –> HasList1 = {WLANbroker=yes, WLANstd=802.11b,
WLANmodel=Cisco62}
  NodeID2 –> HasList2 = {WLANbroker=yes, WLANstd=802.11b,
WLANmodel=Cisco71}
  NodeID3 –>   HasList3 = {WLANbroker=yes,   WLANstd=802.11b,
WLANmodel=Cisco74}

- A wild card shall be represented for example as:

    WLANmodel=Cisco7* –> Cisco71, Cisco74

    And the `getDinaNodes` () method would return:

    [Node2_IP Node3_IP]

storagePointHandler API:

public interface storagePointHandler {

  // New storage point
  public void addStoragePoint(String storagePointID, InetAddress
stpIP, String protocol, int port, String backLevel,
String[] dinaNodeIdList);

  // Recover info
  public InetAddress getIP (String storagePointID);
  public String getProtocol(String storagePointID);
  public int getPort (String storagePointID);
  public String getBackLevel (String storagePointID);
  public String[] getDinaNodeIdList (String storagePointID);

  // Delete
  public void removeStoragePoint(String storagePointID);
}

6.4.1.5. Code Distributor Condition Evaluator

The Code Distributor Condition Evaluator is intended to monitor and evaluate conditions regarding code distribution and maintenance purposes. The conditions regarding code distribution can refer to monitoring events that must trigger some code distribution actions, such as the reception of new service code, etc.

The code maintenance policies are responsible for assuring the maintenance of the service code that is stored at the Code Storage Points. The conditions regarding maintenance can include monitoring code expiration events, reception of new code version, etc.

The Code Distributor Condition Evaluator will run outside the active nodes.

6.4.1.6. Service Invocation Condition Evaluator

This Condition Evaluator is responsible for intercepting service invocation events or signals (e.g., all messages using a specific protocol) in order to trigger a chain of actions that will lead to the execution of a specific Context-Aware Service.

The necessity for this kind of component was a direct consequence of the scenarios considered for the CONTEXT project, involving user mobility and reacting to the emergency traffic.

It was noted that services and users could have different expectations about the events or signals to be captured, in addition to differing scope (geographical, network topological, temporal) of the capturing activity. Therefore, these aspects can be left to be specified during the service customization phase. The appropriate SICEs will be downloaded to the correct nodes or places for their execution. Afterwards, as a result of the conditions of the policies that manage the invocation of a service, the SICEs will be configured specifying all the invocation events or variables to attend and the filters to apply. The information derived from the invocation events received and the application of the Condition Requirements specified should be enough to decide without ambiguity the specific service code to execute.

The Service Invocation Condition Evaluator will run inside the active nodes (EE). The SICEs should be installed and executed in particular nodes where the invocation events and variables must be monitored. It is more efficient to install these components close to the nodes where it is likely that the invocation signals will be received.

The WLAN SICE: This is an example of a generic Service Invocation Listener. It monitors WLAN access networks. The goal of this listener is to notify PBMS when an authorized user enters or leaves the area of connectivity of a WLAN. This information is sent to the PBMS, and in CONTEXT 'Super Mother' scenario, described in Chapter 7, as well as in Reference [2] and [4], service instances are invoked or terminated in edge nodes as required. The list of authorized users is dynamically updated by the PBMS.

6.4.1.7. Service Assurance Condition Evaluator

The Service Assurance Condition Evaluator is the generic name for the possible different Condition Evaluators that are intended to evaluate the different variables that have been decided to determine the quality of the service. There could be variables applicable to all services and others specific for each one.

Once the DMC has received the conditions of the policies related to Service Assurance, it will configure the different Condition Evaluators in the appropriate Active Nodes, where the different assurance parameters have to be monitored. Afterwards, the SACE will obtain these parameters from the running service and will apply the correspondent filters or thresholds. If any parameter falls outside desirable margins, a notification will be sent by the SACE to the DMC in order to communicate the event.

The Service Assurance Condition Evaluator will run inside the active nodes (EE). The SACEs should be installed and executed in particular nodes where these events and variables must be monitored. Service assurance functionality is treated in more detail later in this chapter.

6.4.1.8. Policy and Information Model Repository

The Information Model Repository is the logical storage point for the information about Policies loaded in the system, the network where the CAS is going to be provided (Network Inventory), the services deployed, the components installed, and other information entities (Management Information). Such information contains the status and functionality of every service layer-related entity.

This information is of great importance, for example, when evaluating conditions in order to know whether the appropriate Condition Evaluator is already installed. Another important use of the information contained in this repository could be to decide where a Service Code must be stored depending on the capabilities of the different Storage Points available. In the same manner, it could also be important to decide where a Service must be executed depending on the capabilities of the Active Nodes and the conditions introduced by the user at the customization phase. The Service Creation System will have to be aware of the different Action Consumer components and Condition Evaluator components, and mainly of the different actions and Conditions they are able to enforce and evaluate, in order to compose the Management Policies.

6.4.1.9. Code Storage Points

The Code Storage Points are the physical places where the service code is stored after its distribution and before its execution on the Active Nodes. This means that the requested code will be downloaded from these Storage Points.

The Code Storage Points could also store management system components such as Condition Evaluators and Action Consumers.

6.4.2. Domain-Specific Policies

This section identifies the CONTEXT domain-specific policies that were used in the implementation phase of the Policy-Based Service Management.

The policies are grouped into four functional domains, Service Code Distribution, Service Code Maintenance, Service Code Invocation and Execution, and Service Assurance. These are the main policy domains developed so far, although future implementations might delete these to add new policy types according to practical needs.

The high-level description of the policies exposed in this section follows the format

IF (condition 1) [AND | OR (condition n)]
THEN (*action 1*) [AND (*action n*)]

The Service Management Policies control just the service life cycle, never the logic of the service. In this way, Service Management policies are used by the PBSM components of the system to define the Code Distribution and Code Maintenance of the service as well as the Service Invocation, Service Execution, and Service Assurance.

The policies defined in the different service management functional phases identify the conditions and actions that these policies will contain. It is very important to agree on the specification of the policies used in the PBSMS because all the actions and conditions contained in policies must be implemented in the different components playing the role of Action Consumers and Condition Evaluators.

6.4.2.1. Service Code Distribution Policies

These policies govern the distribution of code on the PBSM System Storage Points components. This code refers to the CAS logic. The main points to notice are:

- They will be processed when new customized service code arrives to distribution component.
- These policies drive the deployment process of needed service.
- They specify the specific configuration, required resources, and criteria for optimum Storage Point selection for the service.

Service Deployment Policies Group example: In this section a high-level example of this type of policy is presented.

Service Code Distribution Policy

(1) 'If (customized service B code is received)
    then (configure distribution of service B code and optimum Storage Point selection parameters)'

Conditions: This policy condition should be based on an event and appears when new service code arrives to the Code Distributor component. This event could contain information needed to identify the code just received. The condition is met for this policy if the code arrived is the one of the Service B. The requirements express this fact applying the requirement type 'Match_value' to the event variables in order to check if the code received is Service B code. The component carried out to monitor the event and evaluate the requirements is the Code Distributor (Condition Evaluator).

Actions: The action implies distributing the Service B code using the particular configuration expressed through the action parameters specified in the policy. These input parameters will specify how the code has to be distributed. This action is enforced by the Code Distributor (Action Consumer).

6.4.2.2. Service Code Maintenance Policies

These policies allow the maintenance of the code installed along the infrastructure to support services. The main points to notify are:

- They will be enforced when maintenance event arise. These events relate to new service version, service expiration, storage points resources under desirable margins, high load of invocation petitions, etc.
- The actions enforced will be service code removal, service code update, service code redistribution (changing number of replicas, Storage Point selection criteria, etc.).

Service Maintenance Policy Group Example: In this section a high-level example of this type of policies is presented.

The high-level policies included in this group could be the following ones:

(1) 'If (new version of customized service B code)
       then (remove old code version of service B from Storage Points)
       & (distribute new service B code)'

(2) 'If (customized service B code expiration date has been reached)
       then (deactivate execution polices for service B)
       & (remove code of service B from Storage Points)'

(3) 'If (The number of invocations for service B is very high)
       then (distribute more service B code replicas to new Storage Points)'

(4) 'If (Resources in Storage Point X are under margins for service B)
       then (remove service B code from Storage Point X)
       & (distribute one replica of service B code to a new Storage Point)'

6.4.2.3. Service Code Invocation and Execution Policies

These policies control the monitoring of variables or events that start the execution of a context-aware service that has been subscribed and customized. They will be enforced when a service is invoked. The invocation signal will be used to deduce the service to execute and the execution parameters associated.

Related to the Invocation and Execution policies and from an informational viewpoint, there are three abstraction levels, related to the invocation of a service and the execution of the respective customized code:

(a) I1: The 'raw' information coming from the system devices (e.g., servers, routers, etc.). This information is input to the SICEs. Usually, this information is at the lowest abstraction level (e.g., at the level of userId, password, raw data measurement).

(b) I2: The information coming from the SICEs to the DM, that is the variables in the notifications that the SICEs send to the DM. This SICE output information is basically the filtered outcome of the information input to the SICEs. It could be at the abstraction level of the input information (i.e., at the level of I1). However, it could be at a higher abstraction level, if in addition to filters, SICEs would be configured with appropriate *mappers*.

(c) I3: The information that identifies (a) which customized service(s) to execute, (b) the initial conditions (run-time arguments), and (c) where to execute the customized service code determined for execution. This information corresponds to the input parameters required by the action undertakers (ACs), which will actually download the code for execution.

This information is specific to the CAS creation system, depending on the particular scheme/convention adopted to name the instances of customised code (it should be stressed that this naming is specific to the CAS creation system and not to the PBMS; ideally PBMS should not restrict the naming of code instances). For instance, customized code could be named after the concatenation of subscriptionId and serviceId (or just subscriptionId if we assume that a subscription contains only one service). But, this naming may not be efficient or suitable at all for a particular CAS, since a subscriber may change this customizations in the context of a subscription or may have a number of customization options. Therefore, a more efficient or suitable (for some CASs) naming would be to name customized code after the concatenation of subscriptionId, serviceId, and customizationId, where the latter indicates the appropriate customization to apply at a given time.

The point is that the abstractions in levels I1 and I3 are different. We find it unrealistic to assume otherwise.

The purpose of the service invocation and code execution control process in our CONTEXT system is to go from abstraction level I1 to I3. That is, based on the 'raw' information (at I1 abstraction level), to determine the triple ⟨subscriptionId, serviceId, customizationId⟩ (or any other tuple that the CAS creation system would adopt for naming CAS code instances), as well as to determine the run-time arguments and the place of execution of the code to be executed.

Broadly speaking, the policy-based operation relying on dynamically defined policies, provides flexibility in transitioning from I1 to I3, thus facilitating service introduction (new CASs) and automated service provisioning (of new customizations). This is the beauty and the strength of our system.

The code invocation and execution domain are not only for deducing the conditions necessitating the execution of a CAS code but also for identifying the code instance to be sent for execution and the associated execution parameters.

Regarding the abstraction levels exposed above, we think that the transition from I1 to I3 is conveyed by the nature of the policy: the association of a specific condition to a specific action.

A condition specifies that if some variables (Condition Objects) adopt some specific values (Condition Requirements) then we assume that the condition is fulfilled. We think that the Condition Objects are expressed at the level I1.

This condition will be associated to an action that will execute some specific piece of code with specific initial conditions. The input parameters of the action relates to the I3 level. Keep in mind that the input parameters values of the action can be 'hard-coded' since policy edition/creation and other functions can adopt its values from some monitored variable defined at policy condition. So, the association between the I1 level and the I3 level is directly derived from the association between the condition (I1) and the action of the policy (I3).

Service Execution Policy Group Example: In this section a high-level example of this type of policies is presented.

Service Code Invocation and Execution Policy

---

(1) 'If (invocation event X is received)
        then (customized service B must be executed)'

Conditions: In the example, the policy condition is based on an event (simple or aggregated variables can be also used). This event is called Invocation_Event_X and appears when a particular invocation signal X is received. This event has associated different variables that can represent different kinds of information that the invocation signal can contain (for instance, a user_id and a password included in some invocation signal). The condition is met for this policy if the event appears and the event variables associated accomplish the requirements expressed in the condition. The component carried out in the policy to listen to the event and evaluate the requirements is a suitable type of *Service Invocation Condition Evaluator* (capable of monitor the event type X).

Actions: The action implies executing the Service B using the particularized execution parameters included in the specification of the action parameters. This action is enforced by the *Code Execution Controller Action Consumer.*

6.4.2.4. Service Assurance Policies

These policies are intended to support and oversee the achievement of QoS levels established for different services. Essentially, they are involved with the arrangement and supervision of the service quality indexes computed from system data, and with the decisions needed for corrective actions if the indexes are not within acceptable limits. Three different types of policies have been recognized in the Assurance Policy Group:

- Service Assurance Initialization Policies
  - Processed when a service start its execution. These policies provide the actions needed to configure the running service (SLO) to export assurance parameters.
- Service Assurance Execution Policies
  - Processed after assurance initialization and during a service execution. Provide the actions to be applied when assurance parameters are within unacceptable margins. These policies can be based on different severity levels, applying different corrective actions depending on them.
- Service Assurance Finalization Policies
  - Processed when a service stops its execution. Provide the actions to be applied in order to stop the assurance activity associated with the stopped service.

It is assumed that the Service Assurance Condition Evaluator and the Service Assurance Action Consumer will be the main components responsible for monitor the conditions and enforce the related actions.

Service Assurance Policy Group Example: In this section high-level examples of these types of policies are presented.

Service Assurance Initialization Policies

(1) 'If (customized service B is running)
    then (configure assurance parameters for service B) & (configure local assurance variables)'

Service Assurance Execution Policies

(2) 'If (level=2) & (parameterA>X) then (Action M)'
(3) 'If (level=2) & (parameterB>Y) then (Action N)'
(4) 'If (level=2) & (parameterC<Z) then (level=1) & (Action K)'
(5) 'If (level=1) & (parameterA>X) then (Action P)'
(6) 'If (level=1) & (parameterD>J) then (Action O)'

Service Assurance Finalization Policies

(7) 'If (customized service B is stopped)
        then (stop assurance parameters evaluation) & (remove local assurance
        parameters)'

6.4.3. Service Assurance

6.4.3.1. Functional Overview

Service Assurance allows diagnosis and correction of CAS's problems during
execution time. We have already touched the subject, and here we take a closer
look at it.

There are two main principles behind the design of CONTEXT's Service
Assurance:

(a) Service Assurance behavior for a certain CAS is controlled directly by
    the CAS author through defining assurance policies. This allows a more
    efficient assurance, as the service author has a deep knowledge about how
    the service works and will probably accomplish this task in the most efficient
    manner.
(b) The Service Assurance design avoids disturbing the Context-Aware Service with
    assurance issues, making assurance as transparent to the CAS as possible (i.e.,
    with minimum intervention in the SLO).

The utilization of policies allows the desired flexibility, allowing configuration at
both deployment time and run time.

Figure 6.6 shows the different components of the Service Assurance System.

The picture shows the following components: ASCE, ServiceMonit, Interface
ServiceMonitMeasure, and AS AC. Furthermore, the assurance system uses several
XSLT files per service customization whose goal will be explained further in the
text.

- ServiceMonit: Service Assurance component, in charge of collecting Service
  Assurance Data and Alarms. There is one instance of ServiceMonit per assured
  instance of CAS.
- Interface ServiceMonitMeasure: Interface that every CAS willing service assurance must implement. It provides methods to export service assurance data.
- Assurance Condition Evaluator (ASCE): In charge of collecting service data and
  alarms and information about the active node, as stated in the monitoring
  messages it receives from the DMC, transforming it and sending processed
  information to the policy engine and to the visualization system.

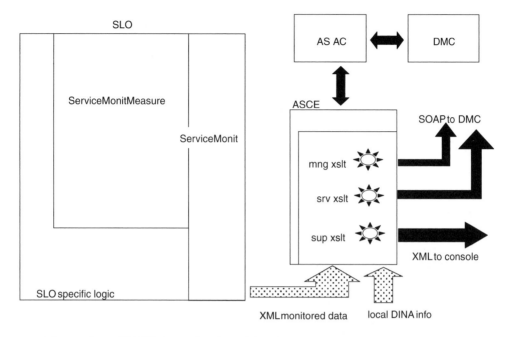

Figure 6.6   CONTEXT Policy-Based Service Management Layer Architecture.

- Assurance Action Consumer (AS AC): It is in charge of doing two tasks:
1. Configuring the ASCE for monitoring the CASs in the active node based on the information received from the PBMS (CodeId of the monitoring XML containing the monitoring parameters to be applied and XSLT's CodeIds).
2. Querying the CD for the URL of those assurance monitoring XMLs and XSLTs. The monitoring XML, updated including involved XSLT's URLs, is passed to the ASCE as active data.

The ASCE uses some XSLT files (service XSLTs and management XSLTs) to transform collected data into a suitable XML to send as an event to the policy engine and to establish thresholds, sending only relevant data. It also uses some other XSLTs (supervision XSLTs) to transform the data into a suitable format for visualization.

Service Assurance Functionality can be split in two main parts, Service Monitoring and Service Management Actions, described below.

6.4.3.2. Functional Decomposition

Service Monitoring: The Assurance Condition Evaluator (ASCE) collects CAS's execution time data and alarms. It also collects some performance parameters about active sessions in the local DINA node. Assurance Policies define which data to collect at a given moment and with which period.

There are two kinds of CAS assurance parameters:

(a) Service specific: (the way to calculate them is up to the service creator).
(b) Standard: (Utilization, Congestion, Reliability, Performance). Although these parameters are common to all services, the way to calculate them must be specified by the service creator, too.

The service must be capable of calculating all of those parameters in the specified way.

Each CAS must run an instance of class ServiceMonit (see 1.2.3.b). This class works as an intermediary between the CAS and the ASCE. When a monitoring message asking for initializing service assurance for a CAS is received in the ASCE, the ServiceMonit instance attached to that CAS uses the interface ServiceMonit-Measure to let the service know that it must begin to calculate assurance parameters. Periodically, and as indicated by further monitoring messages, ServiceMonit will use interface ServiceMonitMeasure to ask the CAS for the values of that parameters.

The way to monitor a CAS (monitoring period, service parameters to be monitored) is defined in monitoring XMLs, linked one to one to assurance policies. These XMLs are defined in the authoring phase, when defining assurance policies, and are used by the ASCE to configure CAS monitoring. The ASCE keeps on monitoring each CAS as specified in the last monitoring message received for that CAS, until a new message is received. Monitoring XMLs are linked one to one to a service XSLT containing the conditions under which an assurance event must be sent to the DMC. While the CAS is being monitored, the ASCE keeps on collecting its assurance data and alarms if any, and if the conditions in the current service XSLT are met, an event is sent to the PBMS.

Thus, the Condition Evaluation functionality is achieved by the utilization of a set of XSLTs. XSLTs are used to transform assurance Data and Alarms, which are initially in the XML format specified by the service creator, to another XML, suitable for sending it as an event to the engine. A XSLT allows thresholds to be established as indicated in monitoring policies so that only relevant pieces of data will be sent. Service Assurance will use three different types of XSLTs. These XSLTs are written by the service creator. This gives flexibility to the system, as the XML format in which the assurance data are exported is not fixed. For a given XML format, XSLTs can be defined in order to fulfill ASCE output requirements.

(a) Service XSLTs: These XSLTs are associated to assurance monitoring policies (there is one XSLT per monitoring policy). The service creator is responsible for defining one service XSLT for each monitoring policy. If conditions in the XSLT are met, the result of applying this XSLT is that an event containing relevant assurance data or alarms is sent to the decision-making component.

(b) Management XSLT: As said before, the ASCE periodically collects information about active sessions running in the local node, in order to allow the service creator to take this info into account in service behavior, using it in the assurance policies defined. There is one Management XSLT per CAS instance, to allow the system to set up thresholds and to send to the policy engine only relevant data. The service creator is responsible for defining this XSLT.

(c) Supervision XSLT: These XSLT is used to transform the XMLs with CAS Assurance Data and alarms. The output is passed to the visualization module (in.html format), to allow human supervision of the executing CAS. The service creator is responsible for defining this XSLT.

Example: The following paragraphs show a simple example to illustrate what has been explained so far about monitoring XMLs and service XSLTs.

Imagine that a new CAS has been started in the active node. For example an instance of Crisis Helper. The CAS author has defined the following policy, stating that when a new instance of CH is executed, the ASCE must start monitoring it as specified in CH_XML_001:

If (a new instance of CH is executed)
then (configureASCE(ReceivedEvent.Service_Id, CH_XML_001))

When the new instance is executed, the policy is evaluated and the ASAC sends the following monitoring message XML to the ASCE:

```
MONITORING MESSAGE (CH_XML_001)
<?xml version="1.0" encoding="UTF-8"?>
<DMCtoCE_Message>
   <ServiceId/>
   <Control>Start</Control>
   <Body>
      <Condition_Object>
        <Event>
            <XMLId>CH_XML_001</XMLId>
            <Event_Type>Assurance_Event</Event_Type>
            <Event_Variable>
                    <Name>Parameter1</Name>
                    <Syntax>String</Syntax>
            </Event_Variable>
            <Event_Variable>
                    <Name>Parameter2</Name>
                    <Syntax>String</Syntax>
```

```
                    </Event_Variable>
                    <Event_Variable>
                              <Name>Parameter3</Name>
                              <Syntax>String</Syntax>
                    </Event_Variable>
        <Service_XSLTId>CH_XSLT_001</Service_XSLTId>
                    <SequenceId/>
                    <SessionId/>
          </Event>
        </Condition_Object>
        <Evaluation_Period>5000</Evaluation_Period>
        <Evaluation_Delay>0</Evaluation_Delay>
</Body>
</DMCtoCE_Message>
```

The ASCE has been configured to start monitoring the CAS. It will evaluate CH's Parameter1, Parameter2, and Parameter3 every 5 seconds.

The service XSLT associated to this monitoring XML is CH_XSLT_001, which is shown as follows.

In the XSLT we can see that the condition specified to send an event to the DMC is that Parameter1 > 600. If the threshold (Parameter1 > 600) is crossed, an event identified by CH_XML_001, and formed as stated in the XSLT, will be sent to the Decision-Making Component in the policy engine.

Note: Some XML lines are too long to fit on the width of the page. To indicate that the line continues to the next row we have used '\' at the end of the row. We hope that this arrangement does not cause any confusion.

```
SERVICE XSLT (CH_XSLT_001)
<?xml version="1.0" encoding="ISO-8859-1"?>
<xsl:stylesheet     version="1.0"
 xmlns:xsl="http://www.w3.org/1999/XSL/Transform">
<xsl:template match="/">
<xsl:if test="Assurance_Data/Specific/Parameter1/Value\
>600">
<CEtoDMC_Message          xmlns:xsi="http://www.w3.org/2001/XMLSchema-
instance"
xsi:noNamespaceSchemaLocation="C:\eclipse\workspace\\
policySquemas\CEtoDMC.xsd">
<Message_Id/>
  <Body>
   <Condition_Object>
        <Event>
          <Event_Id/>
```

```
        <Event_Type>Assurance_Event</Event_Type>
        <Event_Variable>
        <Name>XMLId</Name>
        <Syntax>String</Syntax>
        <Value>CH_XML_001</Value>
    </Event_Variable>
    <Event_Variable>
      <Name>ServiceId</Name>
  <Syntax>String</Syntax>
  <Value>
  <xsl:value-of \
select="Assurance_Data/\
                      Standard/ServiceId"/>
      </Value>
    </Event_Variable>
    <Event_Variable>
      <Name>Parameter1</Name>
      <Syntax>
      <xsl:value-of select=\
      "Assurance_Data/Specific/\
      Parameter1/Type"/>
      </Syntax>
      <Value>
      <xsl:value-of select=\
      "Assurance_Data/Specific/\
       Parameter1/Value"/>
      </Value>
    </Event_Variable>
.........
    <Event_Variable>
      <Name>Performance</Name>
      <Syntax>
      <xsl:value-of select=\
        "Assurance_Data/Standard/\
        Performance/Type"/>
      </Syntax>
      <Value>
       <xsl:value-of select=\
        "Assurance_Data/Standard/\
        Performance/Value"/>
      </Value>
    </Event_Variable>
```

```
        </Event>
      </Condition_Object>
    </Body>
  </CEtoDMC_Message>
</xsl:if>

</xsl:template>
</xsl:stylesheet>
```

If the mentioned event CH_XML_001 has been sent to the DMC, the following policy will be fulfilled for the CH instance:
If (Event Received so that (ReceivedEvent.Event_Type="Assurance_Event" && ReceivedEvent.XMLId="CH_XML_001"))] then (configureASCE(ReceivedEvent.-Service_Id, CH_XML_002))
And the ASCE is configured to monitor the CH instance as specified in CH_XML_002.

Service Management Actions: Assurance Policies can fire actions over the Service Assurance System, in order to configure monitoring, or either over the CAS, to take specific service configurations.

(a) Actions over Service Assurance: The Assurance Action Consumer is the entity responsible for configuring the ASCE with new monitoring XMLs containing monitoring parameters and service XSLTs.
(b) Actions over the CAS: They are taken invoking other Action Consumers.

External Interfaces: As stated previously, service assurance has external interfaces with the policy engine and the Context-Aware Service it is monitoring.

(a) Interfaces with the Policy Engine: The communication with the policy engine is done via two different mechanisms, depending on the direction of the communication.

(DMC -> ASCE) communication
This communication is taken via the Assurance Action Consumer (AS AC). The DMC must invoke the following assurance AC method:

configureASCE(String serviceId, String XMLId, InetAddress nodeIP);

serviceId is the CodeId of the CAS to be monitored. XMLId is the CodeId of the monitoring XML to be applied to that CAS, and nodeIP is the IP of the active node where the ASCE and the CAS to be monitored are running.

(ASCE ->DMC) communication

The communication in the other direction is done via SOAP messaging process. The ASCE is able to send SOAP messages by using the class CientSOAP. This class implements the method send:

public static void send(String xmlstring, String URL);

By calling this method, the ASCE will be able to send an event to the DCM. The content in String xmlstring is sent as a SOAP attachment to the servlet located in the mentioned URL, which is in charge of receiving SOAP messages and passing them to the DMC.

(b) Interfaces with Service Creation & Authoring: Every CAS which needs service assurance must instantiate a service monitor (serviceMonit class), which is part of the Service Assurance system, and is responsible for asking the service for monitoring data at specific periods and for passing the information to the ASCE.

(ASCE -> CAS) communication

CAS services must implement interface context.tid.assurance.service.ServiceMonitMeasure, which grants the ASCE access to service assurance data and alarms. This interface specifies the following methods:

- void setSpecific(String[ ] specificData): Method called by the serviceMonitor (serviceMonit class) to change the current set of exported CAS specific assurance data (standard data are always exported).
- void startMonitoring (): Method called by serviceMonit to let the CAS know that assurance monitoring has begun. The CAS must begin to calculate service parameters.
- String getAssuranceData(): Method called by serviceMonit to fetch assurance data for the current period. Returns a String that contains CAS assurance data in XML format.
- void stopMonitoring (): Method called by serviceMonit to let the service know that assurance monitoring has stopped. The CAS stops calculating assurance parameters.

(CAS -> ASCE) communication

On the other hand, the service uses the serviceMonit's method throwAlarm:
- throwAlarm (String AlarmName): Method used by the service when it detects a service alarm. The alarm is passed to the ASCE in an XML formatted string.

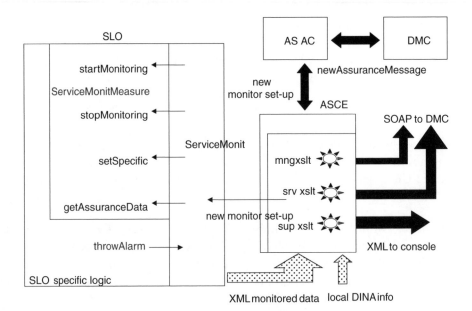

Figure 6.7   Flow Diagram.

The flow diagram is represented in the Figure 6.7
Internal Interfaces: Internal communication between Service Assurance Components takes place basically via temporary files stored in the DINA node, except the communication between the Assurance Action Consumer and the ASCE (The Action Consumer is located outside the DINA node).

(ServiceMonit -> ASCE) communication: ServiceMonit makes available to the ASCE the assurance Data and alarms produced by each CAS
(ASCE -> ServiceMonit) communication: The ASCE configures serviceMonit in order to make the monitoring changes in accordance with the monitoring XML received from the ASAC.
(Assurance AC -> ASCE) communication: The assurance AC sends in an active data packet to the ASCE the required data for configuring it for monitoring. This data contains a XML with monitoring parameters and the URLs of the related XSLTs, if any.

References

1. Fry M, Ghosh A. *Application layer active networking*, HIPPARCH '98 Workshop.
2. Jean K, Yang K, Galis A. *A Policy Based Context-aware Service for Next Generation Networks*, 8th London Communication Symposium, 8–10.9.2003, London, http://www.ee.ucl.ac.uk/lcs/index.html.

3. Galis A, Serrat J, Raz D, Juhola A, Georgatsos P, Serrano JM, Justo J, Marín R, Cohen R, Ahola K, Damilantis T, Vardalachos N, Jean K. *ContextWare Programmable Middleware*, 2nd International Workshop on Managing Ubiquitous Communications and Services (MUCS 2004), Dublin 13–14 December 2004; http://mucs2004.org/.
4. Yang Y, Galis A. *Policy-driven Mobile Agents for Context-aware Service in Next Generation Networks*, MATA 2003- IFIP 5th International Conference on Mobile Agents for Telecommunications, 8–10.10.2003; Marrakech, Morocco; www-rp.lip6.fr/MATA03/.
5. Sygkouna I, Vrontis S, Chantzara M, Anagnostou M, Sykas E. *Context-Aware Services Provisioning on Top of Active Technologies*, IFIP 5th International Conference on Mobile Agents for Telecommunication Applications (MATA 2003), 8–10.10, 2003, Marrakech, Morocco.
6. Nwana HS, Ndumu DT. A brief introduction to software agent technology, Jennings N, Wooldridge M (eds), *Agent Technology Foundations: Applications and Markets*, Springer: Berlin, 1998.
7. Xynogalas S, Chantzara M, Sygkouna I, Vrontis S, Roussaki I, Anagnostou M. Context management for the provision of adaptive services to roaming users. *IEEE Wireless Communications* 2004; **11**(2): 40–47.

7

The Service Execution Environment and Context Delivery

In the previous chapter, we started to outline the architecture of the CONTEXT system. In this chapter, we continue this description and concentrate on the Active Application Layer. We provide a detailed description of the active application network platform DINA, which is the distributed execution environment of the CAS system. We also provide a detailed description of the context delivery system, explaining how context information is made available for the different components of the service.

7.1. A Bird's-Eye View

In the previous chapter, we saw that from the technical point of view the CONTEXT solution comprises three domains, namely: The Service Layer (SL) domain, the Active Applications Layer (AAL) domain, and the IP domain. A different way to consider the various aspects of the same system is depicted in Figure 7.1, where the system is composed of three functional layers, namely the Service Creation layer, the Service Management layer, and the Service Execution layer. The Service Creation layer and the Service Management layer were described in great detail in the previous chapter, and in this chapter we concentrate on the Service Execution layer.

The Service Execution layer is supported by a mixture of active network (AN) nodes and dedicated servers. This layer supports a distributed execution environment in which service code can be executed in a controlled, managed, and efficient way. The motivation for selecting active technology as the main technology to support this distributed service execution environment is, as explained in the first chapters of

---

Fast and Efficient Context-Aware Services Danny Raz, Arto Tapani Juhola,
Joan Serrat-Fernandez, Alex Galis © 2006 John Wiley & Sons, Ltd

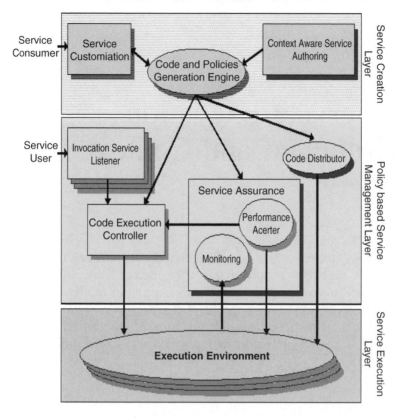

Figure 7.1   CONTEXT Functional Architecture.

this book, due to the advantages that it offers as far as context collection, processing, and storage are concerned.

As described in Chapter 2, many active network platforms could have been chosen as the base for the CONTEX active environment solution. Among all these solutions, the CONTEXT project adopted the DINA programmable platform, which is based on concepts used in the ABLE and ABLE++ systems [1] to provide the programmable context-network functionality. DINA is a programmable middleware which can be attached to different types of network elements and makes them active. In this way, we get a flexible, scalable, and efficient distributed execution environment for the CAS logic, as described later in this chapter. The front end of this executing environment is the APIs provided by DINA to the service code (and to the code of other components of the system). Apart from a unified way to create distributed applications (including support for communication and resource control), these APIs allow the service logic an easy and unified access to local network data and context information, as well as performing actions (such as network-level configurations) as needed. As such, the CONTEXT platform can support context-aware services with a variety of network technologies and applications, in a scalable, safe, and reliable way.

7.2. The Active Platform

The DINA active platform was developed by the CONTEXT project based on concepts and ideas used in the ABLE and ABLE++ systems [1]. In developing DINA, a strong emphasis was put on the missing components needed to create a scalable platform allowing easy deployment of network related context-aware services. This section presents DINA's software architecture, its various software modules, and explains the way in which these modules interact. It also presents the data flow of active services from creation to termination, with emphasis on the interaction with the IP and AN API.

DINA is a modular and scalable software architecture that enables deployment, control, and management of active services (sometimes called sessions or active sessions) over network entities such as routers, WLAN access points, media gateways, and servers that support such services in IP-based networks. In addition to the deployment, control, and management capabilities, DINA provides scalable, platform-independent interfaces that can be used by the active services to manage, control, retrieve information or perform other operations in the local node.

The DINA active platform consists of an Active Engine attached to a Forwarding Element, which can be a router, a WLAN access-point, a media gateway, etc. (see Figure 7.2). The modular design of DINA allows the various logically separated components to be either physically separated or co-located at the same machine. In particular, the Active Engine can reside inside a router on a special card, or at a

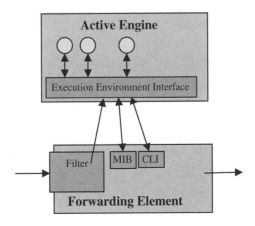

Figure 7.2   DINA Active Platform Environment.

different physical box next to the router. This kind of architecture enables DINA to support different platforms from different vendors using almost identical software components.

In the DINA system, control messages (and possibly the code itself) are sent inbound, that, is using the same infrastructure as the data traffic. In our case, control messages are active packets containing either the code itself, reference to the code, or information exchange between different components of a distributed active application. This choice is very natural in the IP world where all messages are eventually sent in IP packets over the same links. However, in order to implement such an architecture one needs to deploy a mechanism that can filter out the control messages and redirect them to the control plain, this is done in the Forwarding Element.

The Forwarding Element must be able to filter control packets and to send them to the Active Engine. In addition, it must either support standards such as SNMP and MIB or a proprietary but open interface. These interfaces are used to access local information, and to reconfigure the node as required. The information accessed through this interface may concern local resource utilization, load on the different interfaces (when the network element is a router), or information regarding the different call setups (when the network element is a VoIP gateway). Access capabilities may include reconfiguration of the element, changing of parameters and policies, or modifying the QoS parameters or routing tables. Most off-the-shelf IP elements indeed support both filtering and redirection, and can be accessed either *via* SNMP or *via* proprietary CLIs. Some network elements even support more complex access methods based on Java or WEB services technology. The communication between DINA's modules is usually done by UDP transactions and TCP connections, although other methods can be used. In the current implementation, DINA active packets (that should be captured by the filter) are identified as UDP packets with destination ports 3322, 3323 or 3324 (see Subsection 7.2.4). In order to capture such packets, the Forwarding Element should employ a policy-based routing mechanism that allows packets to be filtered and captured according to predefined rules. Using this kind of mechanism, such active packets are captured and are redirected to the Active Engine (actually, as described in the next paragraph, to the Session Broker). The implementation of this filter depends on the platform and the interface with the policy-based routing mechanism, in the specific Forwarding Element.

The heart of the DINA system is the Active Engine. This is a modular software element that offers a controlled, safe, and managed execution environment to the code of the service logic, and other components that perform additional required functions. It is based on a collection of brokers (see Figure 7.3). The Session Broker is the main software component. It handles new control messages, creates and executes the service code, and manages the Active Engine. In addition to the Session Broker, other brokers provide different APIs that allow active services to utilize host

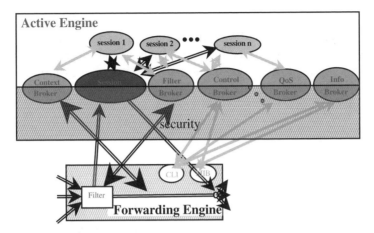

Figure 7.3   DINA Block Diagram Architecture.

information and resources, and to perform operations in the local environment. The concept of using brokers has the following advantages:

- Improving the system security and protecting it against harmful services that may cause (deliberately or accidentally) problems in the active node. The usage of brokers, in addition to other security mechanisms (see Subsection 7.2.5), prevents direct access to sensitive information or resources in the local environment. If a service needs to access such resources, it must use the appropriate broker that can decide, based on its policy, whether to allow the required request or to deny it.
- Enabling advanced access control, based on the services and the required resources or information. In addition to a basic binary access control that determines which services and which users can run services on an active node, each broker can determine its own policy for utilizing its resources.
- Providing atomic platform-independent interfaces that can be used as a building block for Creating Services. Access to resources and information utilization is different from one platform to another. The usage of brokers enables services to access these resources regardless of the platform and the system that hosts the service.
- Scalability and Stability. Each broker provides a set of operations that can be performed by the services using the interface broker API. When new operations need to be added according to the requirements of a new service, this may be done by implementing new brokers that will meet these requirements. In addition, when a broker fails, it impacts only services that use this broker while other services that do not use this broker can continue their work unaffected.

The service logic uses the functionality provided by the brokers via broker interfaces. For each broker, the broker interface is a Java class that can be used by the code implementing the service logic and contains the functionality needed. For example, if the service logic needs to monitor the load on a local interface, it may use the *getLoad* method in the InfBroker class. A detailed description of the different brokers and their interfaces is provided later in this chapter.

7.2.1. The Session Broker

The Session Broker runs the Active Engine and is the core of the active node. It receives and parses control messages (IP packets identified as active packets by the Forwarding Element's filter). If new code implementing part of a service needs to be executed, the Session Broker retrieves the code (if not available already) and executes it in one of the available JVMs. If the control message contains information required by one of the sessions, then this is forwarded by the Session Broker. The Session Broker is an event-driven application that receives events from different communication channels and from an aging mechanism. It communicates with the different services via the Session Broker Interface class instance in each service. As explained above, unlike other brokers, the Session Broker is essential for the execution of any service in the system.

The Session Broker has several fixed communication channels. The external communication channel is used to receive and send active packets. An active node can receive such packets from other active nodes or from the originator host. Active packets are sent to other active nodes according to services requests. If a new code implementing part or a service needs to be executed, the Session Broker retrieves the code (if not available already) and executes it in one of the available JVM's dedicating for active packets. If the control message contains information that is sent to one of the sessions, then the Session Broker sends it to the appropriate session.

The data channel and the admin channel are used to communicate with the local services. The data channel is used to send active payload from the Session Broker to the active session and *vice versa*. Using the admin channel, a service can send control messages to the Session Broker. The following control messages can be sent:

'New': When the Session Broker establishes a new active session, the session enters a 'Pending' state. In this state the service is not yet ready to receive and handle data, and therefore the Session Broker drops any active packet directed to this service. When the Session Broker receives a 'New' control message from an active session, it changes the session state to 'Alive.' Once the session is in this state, it can receive active packets from the Session Broker.

'Refresh': The Session Broker employs an aging mechanism to terminate old and unused active sessions (see detailed description lter in this section). In order to avoid termination due to aging, an active session should refresh itself by sending a 'Refresh' control message from time to time.

'Kill': An active session sends a 'Kill' control message to terminate itself.

The JVM channel is used to communicate with the JVM's that are attached to the Session Broker and used to execute Java-type active services (see Subsection 7.2.2). The Management Broker channel is used to communicate with Management Broker interfaces that may be used by the active services to receive, request, and retrieve information regarding the current status of the Session Broker.

7.2.1.1. Session Database

The Session Broker maintains a database, which contains information about all active services that are currently present at the system. For each service the following information is kept:

- State. This field describes the state of a service and it can be in one of the following states: 'Pending' or 'Alive.'
- Aging. This field contains aging information of a service, namely the service time to live. When this time is expired, the Session Broker should terminate the service. A service can renew its time-to-live by sending a 'Refresh' control message to the Session Broker.
- Sequence number and session ID. The combination of sequence number and the session ID uniquely identifies the service in the active node. These fields are extracted from the DINA header (see Subsection 7.2.4).
- IP address and UDP port. These fields are used in order to send data to a service (via data channel).
- JVM. This field contains information regarding the JVM that runs the service when the active code is written in Java (see Subsection 7.2.2). The Session Broker uses this information to control the service (e.g., terminate it) when this is necessary.
- Security and authorization information. This is information regarding the authorization and access list of the service. More details are provided in Subsection 7.2.5.

7.2.1.2. Aging Mechanism

In order to prevent old, unused services from utilizing the active node resources the Session Broker employs an aging mechanism. The aging mechanism allows the Session Broker to terminate services that are considered to be unused. A service can avoid termination as a result of aging by sending periodic 'Refresh' control messages to the Session Broker (via the admin channel).

In order to fully understand the role of the Session Broker let us consider the life cycle of a part of a service code in a specific DINA node. Such an execution begins when the code of a service (or a reference to a URL containing this code) arrives to the Active Engine. The Session Broker needs to create a local session, and to execute the relevant code. If the active packet contains the actual code then the Session Broker executes this code in one of the available JVM's, and once the code is running it sends a message via the control channel, causing the Session Broker to update the session status to 'Alive.'

Recall that one or more classes in an active packet may be referred by a URL. In this case instead of sending the class in an active packet, one can send a URL for this class. The URL can be an HTTP or an FTP server or a file in the local file system. The URL Class Loader (UCL) communication channel is used by the Session Broker to extract the set of classes that are referred by the URL in the active packet. Once this is done, the code is executed in one of the available JVMs as described above.

During the session life time, the code may receive data packets from other DINA nodes, it may access information via the different brokers, perform local actions via the different brokers, access context information via the Context Broker, or send data to other active session related to the same service in other DINA nodes. All these actions are done via the appropriate channels as described above.

7.2.2. Execution Environment

The Execution Environment (EE) is where the active applications are executed. In order to support different types of active application, more than one EE can be used.

The Java Execution Environment is a set of JVM's (Java Virtual Machines) that are attached to a single Session Broker and run Java-type active services. Each JVM can handle more than one service using the Java thread mechanism, and there may be more than one JVM attached to a single Session Broker. In addition, these JVM's may run on different machines (see Figure 7.4). For each service, the Session Broker can determine which JVM to use according to its internal policy (e.g., finding the JVM with minimum active services, finding the JVM with minimum load).

The Session Broker communicates with the JVM's using its JVM channel (see Subsection 7.2.1). This channel enables the Session Broker to control and maintain the JVM's and allows dynamic addition and removal of JVMs. In this way the system performance can be upgraded as needed, and scalability is maintained.

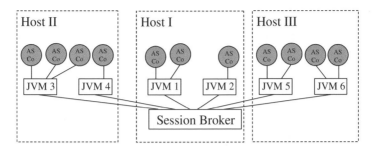

Figure 7.4   Java Execution Environment in DINA.

7.2.3. Management of Active Nodes

As described above, a collection of active platforms provides the distributed environment in which the service logic is executed. Thus, in order to ensure that the service is available, one needs to manage the active nodes. Managing active networks is a very challenging field. One has to verify both that each node is operating properly, and that the network-wide system of active components, such as code invocation and communications, are operating within the designed scope.

In order to manage the active platform, a distributed management application is implemented and run on every managed active node. Active code belonging to this application is sent into each active node. In each node, the management code retrieves local information and controls the active platform. The management application and the active node platform interact using the Management Broker as described below.

The Management Broker Interface provides a management active service that is able to control the active environment. It presents an API to control and retrieve information about the overall utilization of the system resources or about a specific session.

7.2.4. DINA Active Packets

In this subsection, we provide detailed technical information about the structure of the active packets. This information is needed in order to better understand several aspects of the architecture, to explain the IPv6 support, and to be able to analyze the scalability and performance of the system (see next chapter). As explained above, control messages in the CONTEXT system are DINA active packets. These are UDP packets whose payload consists of an ANEP [4] header followed by an active payload.

Since ANEP is a general encapsulation protocol for active networks, a dedicated DINA header extends the ANEP header using the ANEP option field.

IP Header	UDP Header	ANEP Header	ANEP Option: DINA Header	DINA Option: Class Length	Active Payload

Figure 7.5 DINA Packet Format.

DINA active packets are identified as UDP packets with destination port 3322, 3323 or 3324 (Figure 7.5). The different UDP port numbers indicate the type of dissemination used. In the first case we want to sent the packet to the first active node along a path to a given destination; in this case, the destination port of the packet is 3322. In the second case, we want to send the packet to a specific UDP destination. In this case the destination port of the packet is 3323 and only the destination host receives the packet, while all other active (and nonactive) nodes in the route forward the active packet like any IP packet. In the last case we want the packet to arrive to all active nodes along the path to a specific destination. In this case the destination UDP port of the packet is 3324; the packet is captured by every active node in the route of the packet, that is the active packet is forwarded to its destination and each active node on the path creates a local copy.

7.2.4.1. ANEP Header

ANEP is a mechanism for encapsulating active network frames for transmission over different media [4]. The format of the ANEP packet header is shown in Figure 7.6.

The ANEP header fields contents are as follows:

The current version of ANEP, which is described here, and should be set in the 'Version' field, is one. The 'Type ID' field indicates the environment of the message. DINA environment ID is 62. The 'Flags' field indicates what the active node should do if it receives a packet with unrecognized type ID. The node can forward the packet to other nodes (using its standard routing mechanism) or it can discard the packet. The 'Header Len' specifies the length of the ANEP header, including all the options in 32-bit words. The 'Packet Len' specifies the length of the packet, starting from the ANEP header (included), in bytes.

An ANEP header may contain one or more options in the format TLV (Type Length Value). The Type and the Length fields describe the type of the option and its

Version	Flags	Type ID
Header Len		Packet Len
Options		

Figure 7.6 The ANEP Packet Header Format.

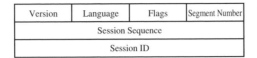

Version	Language	Flags	Segment Number
Session Sequence			
Session ID			

Figure 7.7   The DINA Packet Header Format.

length respectively, and are both 16 bits. The length of the option value varies and depends on the option type.

7.2.4.2. DINA Header

Every ANEP packet that has type ID of 62 (i.e., extended DINA environment) must contain a DINA option. The value of this option follows the format shown in Figure 7.7.

In the DINA packet header, the 'Version' field contains DINA version, which is currently one. The 'Language' field determines the type of the payload and can be zero for data payload or one for Java application payload.

As described in Subsection 7.2.1, in order to capture active packets, an active node, which is not the destination of the packet, should filter these packets using a policy-based routing mechanism. This filtering is done according to criteria that are based on information from the UDP (and IP) header. This mode of operation can lead to a problem when IP fragmentation is used. In this case only the first fragment contains the UDP header and therefore intermediate active nodes will not be able to capture the entire packet.

In order to overcome this problem a mechanism has been employed in the DINA header that enables fragmentation of active packets in the application layer. Using this mechanism every fragment contains IP, UDP, and ANEP (include DINA option) headers (i.e., only the payload is fragmented).

The 'Flags' and the 'Segment Number' fields are used for the purpose of fragmentation. The 'Flags' field indicates whether the current fragment is the last one, while the 'Segment Number' field indicates the reference number of the segment in the packet.

The 'Session Sequence' and the 'Session ID' fields are used to uniquely identify the service in the active environment; thus, no two services may have the same values in these fields. A DINA header may contain one or more options in the format TLV. The length of the options varies and depends on the option type.

7.2.4.3. Class Length and Class URL Option

Active code comprises a set of classes to be executed by the active node. These classes contain a main class that is the first class in the active packet followed by accessory classes.

The Class Length Option and the Class URL Option contain the length of a class or the length of the URL that is presented in the payload. If the payload contains more than one class then the option fields may be concatenated.

7.2.4.4. Active Payload

As mentioned above, the 'Language' field in the DINA header determines the type of payload, which can be one of the following two types:

A new active service. In this case the payload contains an application to be executed by the receiver active node. This application may be followed by a set of parameters for this application. Currently DINA supports Java applications. The application consists of one or more classes in which the first class is the main class to be executed by the active node, while all other classes are used by the main class. Each class can appear explicitly (i.e., the payload contains the bytecode of the class) or implicitly (i.e., the payload contain a reference to the class).

Data for an existing service. In this case the payload contains a set of parameters. When an active node receives such a packet, it forwards the parameters to the related active service, which is already running on the node.

7.2.4.5. IPv6

The system, as described so far, is implemented in IPv4 networks. However, future networks are most likely to use (at least partly) the IPv6 [5] Protocol. Regular applications are not affected by whether the OS works over IPv4 or IPv6 as long as the system calls (i.e., the APIs with the OS) remain the same. However, in our case the application uses IP addresses explicitly. For example, the Session Database contains information regarding the IP address and the UDP port used to send the code invocation command. This information is then used to exchange information between different copies of the application. When the system is used in an IPv6 network, these address fields are different and thus the system must be aware of the change.

Note that the required modifications are technical in nature, and that the detailed implementation and headers definitions as described here are not suitable for IPv6 implementation. In fact, the system has been modified to support both IPv4 and IPv6. This was done by adding a version field and an IP address field before the ANEP header in the active packet, and modifying the Session Broker code and database accordingly.

7.2.5. Security

The concept of active nodes that host active applications makes these nodes vulnerable to malicious services that can use this fact to harm the node and use it for various attacks. To prevent such attacks, DINA uses several security mechanisms that control and constrain active services.

Here, security refers to authenticity and integrity of messages, authenticity of identity and controlling object's access to resources. Secrecy, privacy, and non-repudiation of messages are not considered. More specifically, this section describes how a DINA node is able to verify the source and integrity of an active packet, verify the identity of the encapsulated active code, and apply the correct access control policy as the active code is in the execution environment. Cryptographic mechanisms are utilized to implement these objectives.

7.2.5.1 Relevant Entities

'Context-aware Active Service Provider (CASP)' is the source of active programs. CASP is a logical entity that may encompass a number of nodes, all belonging to the same administrative domain.

'Context-aware Active Network Provider (CANP)' manages the DINA nodes and any supporting elements. The CANP administrator manages policy rules and access control information that are stored in a repository.

'DINA node' is the active node where the active code is executed. DINA nodes retrieve policy rules and access control information from the CANP repository.

'Public Key Infrastructure (PKI)' is a network of Certificate Authorities (CA) and Certificate Repositories. Its function is to manage, store, and deliver X.509 [2,3] certificates and Certificate Revocation Lists (CRL) that enable secure public key based authentication and encryption.

7.2.5.2. Relevant Objects

'Active Service' is the program that is executed in the DINA node. An active service is implemented by active code. Active code either consists of a number of Java classes or may be a native binary application. The active code accesses different brokers of the DINA node. It is created and compiled by CASP.

'Active Code Parameters' are arguments to the active code. Parameters may originate from CASP, and they may be added, removed, or modified by another DINA node.

'X.509 Certificates' encapsulate public keys and associated attributes.

'Permitted Access Control List (PACL)' contains all the access control related attributes. It is managed by the CANP administrator, and is internal to the CANP.

'Static Security Policy File' determines the amount of active node resources that can be used by the active service.

'Java Virtual Machine Security Manager' is a class that allows Java applications to implement and enforce security policies. The class allows an application to identify any requested operation, prior to execution, and determine whether its execution is permitted by the security policies. The application can then allow or disallow the operation.

7.2.5.3. Objectives of Security Functions

The objective of the security functions is to control an active code's access to resources (both information and computation resources) of a DINA node. The dynamic and static security mechanisms prevent the active code's direct access to active node resources (such as I/O, network, and file system resources). These mechanisms force active services to use the different brokers in order to utilize these resources. The static and the dynamic security mechanism complement each other but they differ in their operation. The static system detects security violation before the execution of the active session by parsing the active code. In contrast the dynamic system detects security violations on-line, that is during the execution of the active code. The policy of these security mechanisms is configurable and can be changed using configuration files.

From the CANP point of view, the objectives of authentication and access control mechanisms are to prevent unauthorized use of resources, to prevent the active code from performing harmful functions, and to verify that the active code conforms to rules and policies agreed between the CANP and the CASP. Possible threats to a CANP include hackers and software bugs in active code.

From the CASP point of view, the objective of access control is to assure that the active code does not perform functions that may harm the CANP or any third party. A possible threat are software bugs in active code that could bring the DINA node into unstable state, or flood the network, and thereby cause harm.

7.2.5.4 Architectural Specification

The overall architecture is shown in Figure 7.8.

The following functions are performed by the DINA node:

- Fetch Active Applications code from the Service Provider.
- Get X.509 certificate of the CASP from the PKI Infrastructure.
- Verify the packet and the code authentication and integrity.
- Check if the code is allowed to run in this node. Permissions are handled with PACL (Permitted Access Control List).
- If the PACL is not found in the local repository, fetch it from the CASP's repository.
- Enforce the Access Control Policy instructed by PACL's.
- Execute the Active Code/Application.

The following sections explain the security functions in detail.

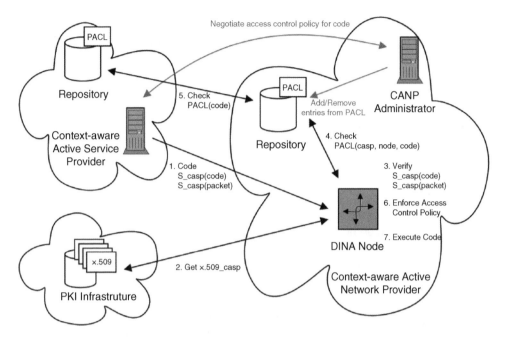

Figure 7.8   Overall Architecture of DINA Security.

7.2.5.5. Access Rights Management

The access control policies within a CANP domain are managed by the CANP administrator.

Before access rights are granted for a new active service, the CASP contacts the CANP administrator and asks the CANP to grant a set of access rights for the active code that implements the active service. The CANP administrator verifies that the active code is safe and valid, and grants the necessary access rights for the code. The access control policies are added to the PACL, and the CANP network is ready for the execution of the active code.

The CANP administrator may also add default policies for unknown active services. The policies may be per CASP, per DINA node, or even global policies with no restrictions.

7.2.5.6. Authentication and Verification of Integrity

General Description: Authentication between the CASP and DINA nodes is based on PKI, X.509 certificates, and digital signatures.

A digital signature provides authentication of the identity and protects the content from unauthorized modifications. The data is signed at two levels:

- Code Signature: The CASP signs the code when it is created, and the DINA node verifies the signature before retrieval of the access control policy for the active code.
- Packet Signature: The sending node signs the whole packet (headers, code, parameters). The sender is either the CASP or an intermediate DINA node. The receiving DINA node verifies the signature upon receipt of the active packet.

Code signing does not introduce significant computing overhead, since the signature is computed only once by the CASP. The code and its signature remain unchanged.

There are two modes to transfer the active code into the DINA node. The code signature is handled somewhat differently in each mode:

1. The active code is sent in the active packet. In this case the code signature is included with the active code in the active packet.
2. The active packet contains a URL-pointer to the active code, and the DINA node uses this URL to fetch the active code. In this case the code signature is included with the URL-pointer in the active packet. The DINA node uses the signature to verify that the fetched code is authentic and unmodified.

Figure 7.9 shows the two authentication modes. Packet signing ensures that the packet cannot be modified in flight without computation of a new signature. Packet

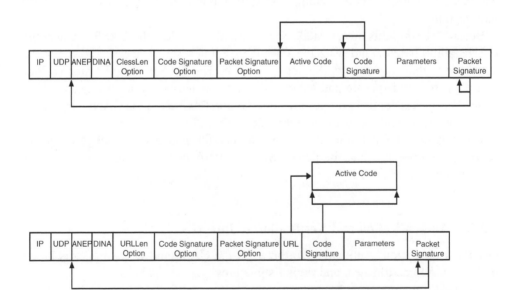

Figure 7.9   Code and Packet Signatures in the DINA Security Mechanism.

signature must be recomputed only when the contents of the packet is changed. Simple forwarding of an active packet does not require recomputation of the packet signature. Both the code signature and the packet signature must be encapsulated with the signer identifying data.

7.2.5.7. Description of Operation

CASP creates a key pair for digital signing purposes, and publishes the public key as a X.509 certificate, which is signed by a certificate authority. There must be chain of trust between the CASP and a root CA that the DINA node trusts. CASP computes a signature for the active code during the creation of the code. CASP builds an active packet with ANEP and DINA headers, the active code (or a URL reference to it), the code signature, and active parameters. CASP computes a packet signature for the active packet, fragments it (if necessary) and sends the packet to the network. Every DINA node that modifies the active packet must recompute the packet signature.

When receiving an active packet, the DINA node verifies the packet signature. First, the source's X.509 certificate is fetched from the PKI. Next, the validity and authenticity of the certificate is verified by forming a chain of trust between the source and the root CA, and by checking the certificate revocation lists (CRL) of the root CA and each intermediate CA. At last, the DINA node uses the source's public key to verify the packet signature. The code signature is checked in a similar way, except that the signer is the CASP (which may differ from the source of the active packet).

Low-level functions such as fragmentation and reassembly of an active packet must not require the recomputation of the packet signature. Thus, the packet signature cannot cover all fields of the ANEP and DINA headers. Rather, a pseudo-header must be used in the signature computation.

7.2.5.8. Replay Protection

An active packet contains a unique identifier that distinguishes two packets from each other. The DINA node maintains state information about the identifiers of received packets.

A packet sequence number provides sufficient protection. The source node stores the sequence number of the last packet it has sent, and the destination node stores the sequence number of the last received active packet that was valid. Upon receipt of a new packet the sequence number of the packet is compared with the stored sequence number. The packet sequence number is considered valid if it is larger (taking rollover into account) than the stored sequence number.

An alternative method is to use daytime as the packet sequence. In this way the source node does not have to store the sequence number. However, this method is sensitive to clock adjustment at the source node.

7.2.5.9. Identification of Active Code and Active Service

The active code that implements an active service is either a binary application or a set of Java bytecode classes. Furthermore, the code may be customized in CASP, that is the active service consists of a base code and customized code.

In the case of Java active services, the service consists of base classes and customized classes. The active service can be identified by the code signatures of the base classes.

7.2.5.10. Access Control

Permitted Access Control List (PACL): All access control-related information is stored in the PACL (in the CANP repository).

The PACL contains mappings from a set of parameters to a set of access control policy rules. The parameters needed to identify a set of policy rules are:

1. Active Code identity.
2. The source of the code (CASP): The CANP can define different access control policies for different CASPs.
3. DINA node: The CANP can define policies based on the attributes of the EE (e.g., location of DINA node in the network, amount of computing resources, etc.)

Parameters may contain wildcards, thus allowing default policies (per CASP for example). The PACL is accessed in the order of parameter accuracy, that is policy rules for a specific (code, source, node)-triple override the default policies.

The CASP may also maintain a PACL, which contains the CASP's preferred access control policies for each of the CASP's active codes. The CANP may utilize this PACL to retrieve access control policies for active code that is unknown but originates from a trusted CASP. This form of PACL chaining allows the CASP to ensure that the execution of the active code does not cause significant harm even if the code does not behave as designed.

The access control policy rules define what functions the active code is allowed to perform in the DINA EE. For example, an active service may be allowed to call certain Java methods, connect to some remote servers, or send data via network at a specified maximum rate.

7.2.5.11. Static Access Control Enforcement

The static security mechanism parses the active code and finds security violations before the code executes. In particular, it detects if the active code tries to access the system resources directly (i.e., without using the different brokers). This security system is employed in the session broker according to predefined security policy that can be dynamically changed using a static security configuration file. This configuration file identifies the resources that can be used by the active service.

In the Java EE, where the active code is a set of Java classes, the static security system checks each one of the classes in the active payload and extracts the instances that are used by these classes. If an instance is not authorized by the static security policy file then the static security system denies execution of the service and drops the code.

Static access control must be extended such that the access control policy from the PACL can be used as input when applying the static access control to the active code.

7.2.5.12. Dynamic Access Control Enforcement

The dynamic security system detects potential security violations during execution of the active code. In the Java EE, the implementation uses the Security Manager mechanism. Each, time a system resource is requested, the security manager checks if the caller is one of the brokers and, if so, it accepts the operation, otherwise the operation is rejected.

Altogether, DINA Security mechanism allows the provision of programmable services with a minimal risk in terms of security. One has to note that the computational cost and the added complexity to the system due to these security enforcement are not negligible and a further study is in place regarding the trade-off between security and efficiency.

7.2.6. The IP-Related Brokers

We describe here the various brokers that allow the service logic to retrieve local information and to perform control actions, as needed. Recall that each network element can be supported by different brokers depending on the required functionality. We start with IP-related brokers.

7.2.6.1. Information Broker

The Information Broker Interface provides services to retrieve local IP-related information such as the name and IP address of routers connected to the local LANs and general MIB objects in the active host.

```
public class infoBrokerInterface
{
public infoBrokerInterface();
public String snmpGet(String s);
public String [] snmpGet(String [] s);
public String snmpGetNext(String s);
public String [] snmpGetNext(String [] s);
public int getNumIf();
public String getRouterName();
public String [] getIpAddrs(int ifnum);
public String [] getIpMask(int ifnum);
public int getIfNumber(String ipaddr);
public String getNextHopAddr(String dest);
public int getNextHopIf(String dest);
public oat getLoad(int ifNum);
public int getStatus(int ifNum);
public String [] getActiveNeighborsAddrs();
public String [] getActiveNeighborsAddrs(int ifNum);
public boolean isLocalLoopback(String addr);
public boolean isLocalLoopback(int ifnum);
public String [] getDestAddrs(int ifnum);
}
```

The Information Broker retrieves information from the active platform using an SNMP client. In addition, it maintains a cache in order to reduce the volume of queries.

7.2.6.2. Control Broker

The Control Broker Interface enables active services to control and configure the routing tables and VPN connections. It is possible to install two kinds of routes into the routing tables: temporary and permanent. Configuration of temporary routes requires one extra parameter. VPN connections can be established also in same manner. If timeToLive is defined as 0, the tunnel is permanent.

```
public class controlBrokerInterface
{
public static int ADD_ROUTE = 0;
public static int DEL_ROUTE = 1;

public static int VPN_IPIP = 0;
public static int VPN_GRE = 1;
public static int VPN_IPSEC = 2;
```

```
public static int START_VPN= 0;
public static int STOP_VPN= 1;
public static int SHOW_VPN= 2;

public controlBrokerInterface();
public String setRoute(int cmd, inetAddr net, inetAddr netmask,
inetAddr gw, String if);
public String setRoute(int cmd, inetAddr net, inetAddr netmask,
String if);
public String setRoute(int cmd, inetAddr net, inetAddr netmask,
inetAddr gw);
public String addTempRoute(int timeTolive, inetAddr net, inetAddr
netmask, inetAddr gw, String if);
public String addTempRoute(int timeTolive, inetAddr net, inetAddr
netmask, String if);
public String addTempRoute(int timeTolive, inetAddr net, inetAddr
netmask, inetAddr gw);
public String getRoutingTable();
public String setVPN(int timeToLive, int tunnelType, int cmd,
inetAddr destIP, inetAddr destNet, inetAddr netmask);
public String setVPN(int tunnelType, int cmd, inetAddr destIP,
inetAddr destNet, inetAddr netmask);
public String setVPN(int cmd, inetAddr destIP, inetAddr destNet,
inetAddr netmask);
public void close();
public getLastError();
}
```

7.2.6.3. Network Broker

The Network Broker Interface provides active services with an interface to basic communication services such as TCP connections, UDP transactions, and connection with other active services.

```
public class networkBrokerInterface{
public static final byte SOCKET_UDP= 17;
public static final byte SOCKET_TCP= 6;
public networkBrokerInterface();
public int socket(byte type);
public int setSocketTimeout(int sd, int timeout);
public int connect(int sd, InetAddress dest, short port);
public int bind(int sd, short port);
```

```
public int accept(int sd);
public int send(int sd, byte[] buf, int length);
public int sendto(int sd, byte[] buf, int length, InetAddress addr,
short port);
public int receive(int sd, byte[] buf, int len);
public int receivefrom(int sd, byte[] buf, int len, InetAddress
addr, short port);
public int sendUDPpacket(byte[] buf, int length, InetAddress
dest, short port);
public int close (int sd);
public String getLastError ();
}
```

7.2.6.4. Filter Broker

The Filter Broker Interface provides ways to control a router to filter traffic.

```
public class filter BrokerInterface{
public String getDataFlow()
public String addRawIPTableRule(String rule, int durationTime)
public String addRawIPTableRule(String rule)
public String getAllRules()
public String addIPTableGetFlowRule(String chain, String filter,
boolean withHeaderData, int durationTime)
public String addIPTableGetFlowRule(String chain, String filter,
boolean withHeaderData)
public String addIPTableGetPartialFlowRule(String chain, String
filter, boolean withHeaderData, int partSize, int durationTime)
public String addIPTableGetPartialFlowRule(String chain, String
filter, boolean withHeaderData, int partSize)
public String addIPTableGetProbabilisticFlowRule(String chain,
String filter, boolean withHeaderData, int rate, int durationTime)
public String
addIPTableGetProbabilisticPartialFlowRule(String chain, String
filter, boolean withHeaderData, int partSize, int rate, int
durationTime)
public String addIPTableAcceptRule(String chain, String filter,
int durationTime)
public String addIPTableAcceptRule(String chain, String filter)
public void removeIPTableRule(String handle)
public void refreshIPTableRule(String handle)
public void resetRuleCounter()
public int getStatistics(String handle)
```

```
public String getRule(String handle)
public String getTypeofRouter()
}
```

While using the same API, the implementation of the API for different network elements (for example a LINUX machine or a commercial CISCO router) is very different. Thus while the front end (the API to the code) remains the same, different code needs to be implemented to support different network elements. This part of the broker code is sometimes referred to as a 'wrapper.'

7.2.6.5. QoS Broker

The QoS Broker Interface provides active services with the ability to configure and manage the routers' network interfaces in order to support QoS functionality. The QoS configurations refer to the Differentiated Services (DiffServ) architecture.

The QoSBrokerInterface API provides to the service developer a set of methods that can be classified into four main categories:

- Router identification: The methods of this category can be used to identify the type of the current router (core, edge-1, edge-2 or out of source-destination path).
- Interfaces identification: The methods of this category can be used to discover the network interface that should be configured.
- Routing: These methods can be used to guide the active packets towards the source or the destination host.
- Configuration: These methods can be used to install DiffServ queuing disciplines/ classes/filters in order to set up classifiers, policers, markers, etc.

```
public class QoSBrokerInterface{
public QoSBrokerInterface(InetAddress src, InetAddress dst, int port);
public QoSBrokerInterface(int port);
public void setSource(InetAddress src);
public void setDestination(InetAddress dst);
public boolean resetInterfaceDiffServ(String iface);
public boolean resetAllDiffServ();
public boolean initINGRESS(String ingress);
public boolean resetINGRESS(String ingress);
public boolean installDSMARK(String egress);
public boolean resetEgress(String egress);
public boolean isLinuxController();
public boolean isEdge1();
```

```
public boolean isEdge2();
public boolean isCore();
public String getIngress();
public String getEgress();
public String getNextHopFor(InetAddress target);
public boolean installCBQ(String egress, double bandwidth);
public boolean setAFxClass(int AF_x, String egress, double band-
width, double rate, int prio, boolean bounded, int DPs, int
default_DP);
public boolean setAFxyClass(int AF_x, int AF_y, byte DSCP_byte,
String egress, int limit, int min_limit, int max_limit, int burst,
double bandwidth, double probability, int prio);
public boolean setBEClass(String egress, double bandwidth, double
be_rate, int limit, int min_limit, int max_limit, int be_burst,
double probability);
public boolean setEFClass(String egress, double ef_rate, int
ef_burst, double ef_mtu, int ef_limit);
public boolean resetAccessLists();
public int createAccessList(int accesslist_id, String protocol,
String source_ip, String destination_ip, int source_port, int
destination_port);
public void removeAccessList(int access_list_id);
public int addMarker(int index, int accesslist_id, byte TOS,
String egress, String ingress, int prio);
public boolean removeMarker(int index);
public boolean removeMarker(int index, int accesslist_id, byte
TOS, String egress, String ingress, String prio);
public int addMarkerPolicer(int index, String egress, String pro-
tocol, String src, String dst, int sport, int dport, int rate, int
burst, int tos, int prio, String policy);
public int addMarkerPolicerCISCO(int index, int accesslist_id,
int rate, int min_burst, int max_burst, int conform_dscp, int
exceed_dscp, int violate_dscp, String egress);
public boolean removeMarkerPolicer(int index);
public boolean removeMarkerPolicerCISCO(int index);
public boolean removeMarkerPolicer(int index, String egress, int
prio);
public String monitorQdiscs(String iface);
public String monitorClasses(String iface);
public String monitorFilters(String iface);
public void finish();
}
```

7.2.7. VoIP Support: the SIP Broker

The SIP Broker allows service logic access to a SIP software package. The broker enables the active services to control and manage SIP entities such as proxy servers and user agents. The SIP Broker provides the SLOs the ability to obtain information from the SIP softswitch and to control aspects of call control during a crisis situation.

The SIP broker performs three primary functions. The first is to provide information about the SIP softswitch and the SIP users. The SIP softswitch is the software that controls call admission, control, and signaling using the Session Initiation Protocol (SIP). The second function is to terminate calls. The third function is to delegate call admission to the SLOs.

```
public class SipBrokerInterface {
//User status
public static int USER_ACTIVE = 0;
public static int USER_INACTIVE = 1;
public static int USER_DISCONNECTED = 2;
public SipBrokerInterface();
public int userStatus(SIPUrl sipUserAddr);
public InetAddress getUserIP(SIPUrl SIPUserAddr);
public int maxSessions();
public int totalSessions();
public String[] getProxyServers();
public String[] sessionStateInfo(String sessionID);
public Hashtable sessionStateInfo(String SIPProxyID);
public boolean terminateSession(String sessionID, SIPUrl Caller-
Addr, SIPUrl CalleeAddr);
public boolean terminateAllSessions(String SIPProxyID);
public void setCallAcceptanceDiverter (String SIPProxyID SIP-
CallDiverter diverter);
public void applyDiverter (String SIPProxyID boolean apply);
}
```

As previously explained, the implementation of the API is network element specific. In this case it depends on the type of the SIP softswitch in use, and a different 'wrapper' should be implemented for each different form of SIP software.

7.2.8. Wireless Support: The WLAN Broker

The WLAN Broker provides active services with the ability to control and manage WLAN Access Points and the Wireless Network. Recall that a VLAN

(Virtual LAN) can be regarded as a group of devices on different physical LAN segments, which can communicate with each other as if they were all on the same physical LAN segment. In other words, a VLAN can be thought of as a broadcast domain that exists within a defined set of devices. A VLAN consists of a number of end systems, either hosts or network equipment, connected by a single bridging domain.

Let us now consider the configuration of a WLAN to constitute a VLAN. The WLAN Access Points are Layer 2 devices. At this level, in order to provide VLAN, we have to encapsulate the packets following the 802.1Q extension of the 802.1D standard. The 802.1Q defines the architecture for Virtual LANs and services provided therein.

According to the previous discussion, we can say that in order to allow a WLAN be part of a VLAN we need to activate the 802.1Q tag awareness in the AP of this WLAN. As depicted in Figure 7.10, the activation consists of a binding between an existing VLAN and a particular Service Set. Service Sets are associated to VLANs in a one-to-one basis. Simultaneously, a Service Set, which is a logical coverage area, is bound to an Access List, which will contain the MAC Addresses of all the users that are allowed to connect to this Service Set. Moreover, a given WLAN AP can also have Service Sets that are not associated to any VLAN. Frames from the wired network with destination to clients belonging to different VLANs are transmitted by the AP to different Service Sets. Only clients associated with a particular Service Set can receive those packets that belong to a particular VLAN. Conversely, packets coming from clients accessing via the WLAN are 802.1Q tagged before they are forwarded onto the wired network.

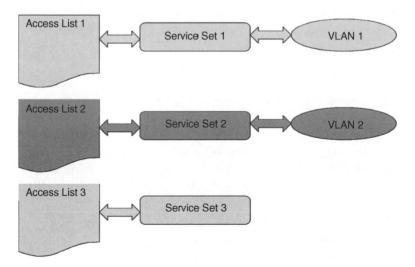

Figure 7.10   Association of VLANs, Service Sets, and Access Lists in a WLAN Access Point.

QoS in a WLAN AP can be assigned in two ways. First, it can be assigned globally to the VLAN. This means that the flows of all users assigned to this VLAN will have the same QoS. On the other hand, we can also assign QoS to individual users whether or not they belong to any of the existing VLANs.

In order to provide a service with a given quality, it is necessary to keep a set of performance values within given thresholds. In other words, there are some parameters that must be configured and monitored in the network in order to assure the QoS. In packet networks these parameters are mainly delay, bandwidth, jitter, and packet loss. In this strict sense, an AP of the 802.11a,b standard has no means to provide QoS. The AP can only establish different packet priorities based on one of the following three mechanisms: DSCP value (Differentiated Services Code Point), client identification, or the priority value on the 802.1Q / 802.1p tag.

All packets tagged with the same priority will be treated in the same way, but will be given precedence over packets tagged with a lower priority. However, this priority does not assure the total amount of bandwidth that they will share or even how much bandwidth they will be able to use. The amount of bandwidth that a given class shares is not guaranteed because it depends on the distribution of traffic among the different traffic classes.

Finally, the interface adopts the following structure:

```
public class WLANBrokerInterface
{
public WLANBrokerInterface(InetAddr APAddr, String confPasswd);
public boolean isUserAssociated(String ClientIPAddr);
public String[] getAssociatedUserAddrs();
public String getUserMAC (String ClientIPAddr);
public String getUserIP (String ClientMACAddr);
public boolean createSS(String SSID, String AutType, String
AccessListID, String MaxAssoc);
public boolean removeSS(String SSID);
public String getClientSSID(String ClientMACAddr);
public boolean establishVLAN (String VLANId, String SSID);
public boolean removeVLAN (String VLANId, String SSID);
public boolean addAccessList (String AccessListID, String permi-
tion, String MACAddr);
public String[] getAccessList ();
public boolean removeAccessList (String AccessListID);
public String snmpGet (String community, String host, String oid);
public oat getLoad();
public oat getLoadPerConnectionfromAP(String ClientMACAddr)
public oat getLoadPerConnectionfromClient (String ClientMA-
CAddr);
```

```
public int getSignalQuality (String MACAddress);
public int getInterfaceStatus (int ifNum);
public boolean setInterfaceStatus (int ifNum, boolean status);
public boolean setChannel(String ChannelNumber, int ifNum);
public boolean isChannelActive(int ChannelNumber, int ifNum);
public boolean setMACFilter (String FilterName, String[] MACAd-
dress, String AccessListID);
public boolean setVLANFilter (String FilterName, String VLANId);
public boolean removeFilter (String FilterName);
public boolean setQoSPolicy (String PolicyName, String Filter-
Name, String Priority);
public boolean removeQoSPolicy (String PolicyName);
public boolean bindQoS (int ifNum, String direction, String Pol-
icyName);
public boolean unbindQoS (int ifNum, String direction, String
PolicyName);
}
```

7.3. Context Delivery System

Context-aware services need a constant access to information about their environment to be able to adapt to it. To support such services there is a need to collect, aggregate, and disseminate context information. This process is often termed *context management* in the literature. Context information is derived from diverse information sources spread over the network. A fundamental requirement then is to collect raw data from thousands of diverse sources, process the data into meaningful context information, either simple or complex, and disseminate the information to applications located at different network locations.

In order to describe the mechanism for the acquisition and distribution of context information, we concentrate on the information providers (these are entities that provide the context information to the system) and information consumers (these are entities that use the information like the service logic). This abstraction allows us to use publish/subscribe APIs relieving the producers and consumers from the details of the underlying dissemination mechanism. Data producers generate data and publish it, data consumers subscribe to data, and it is the task of the dissemination mechanism to ensure that context information travels efficiently from publisher to subscriber. Note that a very similar description was used in Chapter 4 to describe Network Context Information. In this section, we describe the actual system implemented as part of the CONTEXT project, where the Context Mediators are implemented by DINA Context Brokers.

The Context Information Dissemination System (*CIDS*) is the system that handles the actual acquisition and dissemination of the information. The *CIDS* takes into account any special requirements such as: timely distribution of context to a large set of consumers; dealing effectively with the asymmetry that arises from the disproportion between the number of producers and consumers; dissemination of frequent context updates due to the volatile nature of context information in highly dynamic environments; and ensuring the system's ability to evolve and scale gracefully facilitating the seamless integration of new information services without affecting the existing system. By combining known data delivery techniques such as unicast and multicast, the performance and scalability of the dissemination mechanism can be greatly enhanced. In the *CIDS*, in addition to context producers and context consumers, we also have context brokers, which facilitate the communication between consumers and producers.

7.3.1. Functional Overview

The producers of context information include all the Context Information Sources (CISs) that are attached to specific active nodes and provide raw context information, such as: information from DINA Brokers (e.g., local MIB variables), information from GPRS and WLAN Wrappers (e.g., on the corresponding coverage), information from other wrappers (e.g., weather sensors), and various other information provided by *Context Providing Applications* (e.g., user agenda). In addition to these context sources that provide raw context information, producers also include applications that produce complex context information after collecting and aggregating elementary context.

The consumers are the SLOs, the instances of the service logic. During the operation phase, each SLO issues requests for acquisition of context information in order to adapt to the imminent context changes and proceed with the enforcement of the appropriate actions.

The CIDS Brokers, namely the Context Brokers, act as third-party players between producers and consumers. They accept requests from SLOs, collect context information from producers, and efficiently disseminate the information from producers to consumers or other Context Brokers, as dictated by a dissemination scheme. In this sense, Context Brokers realize the linkage between producers and consumers. Each Context Broker provides two different types of API: the Producer APIs and the Consumer APIs. The *Producer APIs* include the interfaces that enable context producers to publish the information they provide, either raw or complex, so that the Context Brokers will then be able to access it. They also include the interfaces that enable the context producers to deliver the information that they have produced. In this sense, the heterogeneity of the different context sources is hidden and each time a modification of the context sources takes place, only the local

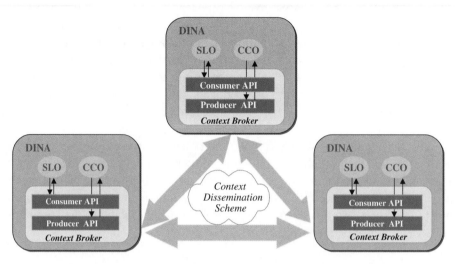

Figure 7.11   Context Information Dissemination System (CIDS).

Context Broker needs to be updated with the new interfaces to interact with them. This process is transparent to the SLO. The *Consumer APIs*, on the other hand, enable an SLO to access context information either through pull requests or notification events.

In the case of complex context information, special Active Applications called Context Computation Objects (CCO) collect raw context information provided by the available context sources, either local or remote, aggregate or filter it, and produce complex context information; thus, CCOs represent CISs of complex context. They are deployed (permanently or on demand) in the DINA EE.

The Context Dissemination Scheme (*CDS*) ensures the efficient and scalable distribution of context information among the different players of the CIDS (Figure 7.11). A variety of data delivery mechanisms exist. Their selection is guided by performance criteria, such as the efficiency achieved in the use of communication resources, scalability enhancements, or responsiveness in terms of access latency experienced by the SLOs. In this sense, different delivery options are exploited. A system based solely on querying – pull mechanism – would suffer from scalability problems. The *server Context Broker* would have to be constantly interrupted to deal with pull requests and could easily become a scalability bottleneck in systems with large *client Context Broker* populations. This problem could be avoided by allowing the information to selectively flow to interested clients – push mechanism – instead of requiring the client to read information periodically. Furthermore, given the similar nature of many context queries, 1-to-N communication can amortize much of the overhead of sending context data to multiple clients. In this framework the consistency of delivered context is also envisaged by enriching the delivery mechanisms with a means to deal with the aging of the context data.

7.3.2. Functional Decomposition

The *CIDS* is implemented by the Context Brokers in the DINA system. Any code that needs to access context information uses the Context Broker interface. Using this API, CCOs can register and supply their context information, and SLOs can request specific context information or to be notified when the value of such an information item changes.

Thus, the requests that are addressed by the Context Broker could be either context queries or context events (pull or push mechanism, respectively). Additionally, there are two options that are offered to the CCOs for providing their information: They could provide their context upon request or they could delegate the Context Broker to serve the submitted requests, which in this case stores the updated values.

Services' SLOs and CCOs forward requests for context information to their local Context Broker, although they may acquire context directly from other local DINA Brokers, if the developer considers this to be the best option. The Context Broker provides a uniform way to access context information, thus hiding the complexity of the context information and its retrieval. Well-defined APIs for retrieving context information facilitate the realization of an automated service creation process.

Context Broker functionality is logically divided amongst the components depicted in Figure 7.12. These components are described in the following subsections.

A common description of context information is required so that both context producers and context consumers can have unified access to the API. For that

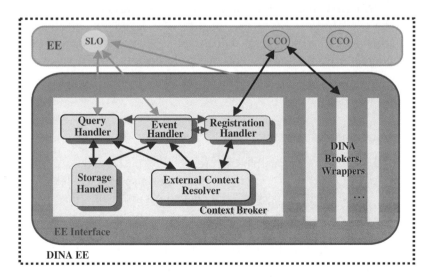

Figure 7.12 Context Broker Architecture.

purpose we use the Context Modules that have been introduced in the previous chapter. Context Modules are XML-based descriptions of context information. A Context Module consists of three parts: the context name, the input parameters, and the output parameters. The definition of a Context Module per context information type ensures that context consumers and producers have a common description of the input and output parameters and this facilitates the efficient contextual information exchange. A context consumer, requesting specific context information, provides the name of the context module, the input parameters (names and values), and the output parameter names it wishes to acquire, and receives the values corresponding to the requested output parameters. A context producer, registering context information, provides the name of the context module and the values of the identification parameters (these are a subset of the Context Module input parameters; the values of the remainder are specified by the context consumer). Context Module XML descriptions are stored in the *Context Module repository*. Regarding the naming of the context modules, the following model is introduced. Each name has two parts: $\langle xxx.xxxx.xxx \rangle @ \langle pppp \rangle$. The first part ($\langle xxx.xxxx.xxx \rangle$) characterizes the context information and can be in a dotted form, in accordance with the logical containment hierarchy of the entity-based modeling. The second part ($\langle pppp \rangle$) denotes the producer of this piece of context information.

7.3.2.1. Query Handler

The *Query Handler* is responsible for resolving 'pull' context requests issued by SLOs. 'Pull' context requests correspond to the retrieval of the output value of a context module, based on the input parameters. The requested context information may either be available locally or in a remote CIS. In this perspective, once the Query Handler receives a context request, it contacts the local CISs. If the requested context information is not available locally, the External Context Resolver is triggered, in order to initiate the mechanism for contacting the remote CISs.

7.3.2.2. Event Handler

The *Event Handler* is responsible for resolving 'push' requests for context information. An SLO may register to be notified when a specific context information data has a specific value or a specific delta change occurs in its value, or to receive periodical updates on the value of a context item in specified time periods. These requests require an asynchronous means of communication. The Event Handler implements the APIs that allow SLOs to subscribe in order to receive a context event. As with the Query Handler, the Event Handler contacts the responsible CISs (local or remote) to acquire the relevant data.

7.3.2.3. Registration Handler

The *Registration Handler* provides a mechanism that enables the registration of context data items in the system. The Registration Handler offers the interface used by context producers to publish the context information that they provide, hence informing the appropriate Context Broker of its existence. In this sense, we foresee a data-centric approach, where the context producers supply data matching certain properties, rather than data from particular sources. Such an approach facilitates evaluation of the available context producers in order to select the one that best serves context consumer's needs.

7.3.2.4. Storage Handler

Two options have been considered for storing and retrieving context information. These options are differentiated based on whether a context producer retains the task of providing the data that it generates itself or if it delegates the task of answering the context request queries to third party. The *Storage Handler* accommodates the latter case, by providing the corresponding interface for delegation. Thus, it is responsible for retrieving the context information from these CISs, storing it, updating it, and disseminating it as needed.

However, the aforementioned option presupposes that the context information is stored before being requested from a context consumer. Such a solution could result in wasteful resource consumption (such as network resources, storage resources, time required) in case it is used as a standalone solution. Therefore, the context producers that will delegate the storage of their data need to be carefully selected based on various parameters.

7.3.2.5. External Context Resolver

If the Context Broker cannot find the requested context information locally then the *External Context Resolver* is triggered. In a distributed context provisioning system a mechanism is required to address the issue of retrieving context from remote sources. Different delivery mechanisms that ensure the efficient dissemination of context information among different Context Brokers could be considered. In a distributed information system, the search phase is in many cases the most time-consuming one. The approach adopted in CONTEXT that exploits active networks is a flooding of the registration information to all the Context Brokers; whenever a new CIS registers with the system through a specific Context Broker, the information about the registered item is broadcast to all Context brokers in the system. In this sense, the External Context Resolver is responsible for disseminating the location of a new CCO registered with a specific Context Broker to the rest of the Context Brokers. After having detected the location of a remote CIS that provides the requested context information, the External Context Resolver establishes a UDP/TCP connection to the appropriate Context Broker to retrieve the information. Moreover, the potential of a Context Broker periodically to receive context information provided by other Context Brokers without explicit requests, as in demand-driven access, could occasionally prove cost-effective, and thus a prefetching mechanism is also supported. Finally, different delivery options based on broadcast/multicast to save resources when similar needs of consumers come from different points of the network are under consideration.

7.3.3. Context Broker Interfaces

The Context Broker offers the following APIs to the SLOs and CCOs.

```
public class context BrokerInterface{
public String[] register_new_context_object (String context_-
name, String reg_Id, boolean isDelegated, String[] reg_parms,
String[] parms)
Public String register_external_new_context_object (String object_-
name, String reg_Id, boolean isDelegated, String[] reg_parms)
public void delete_reg_context_object (String reg_Id)
public String delete_external_context_object (String reg_Id)
public String[] get_current_value (String context_name, String[]
parms)
public String[] subscribe (String context_name, String notifica-
tion_Id, String[] parms)
public void delete_subscription (String notification_Id):
public void supply_context_value (String context_name, String[]
context_value, time timestamp)
public void deliver_notification (String notification_Id, String
requestor_ipaddress, String notification_value, time timestamp)
private Registration_Object get_registration_info (String con-
text_name, String parms)
private delegate_new_context_object (String context_name, String
deleg_Id, String del_parms)
private void delete_delegate_context_object (String deleg_Id)
private void send_packet_register (String object_name, String[]
reg_parms, String port)
private void send_packet_delete_register(String reg_Id)
}
```

The design and implementation of the Context Information Dissemination System, using the Context Broker in the DINA system, and the context item definition, results in a modular scalable system that can efficiently support CASs in a distributed execution environment.

7.4. Conclusions

In this chapter, we described the distributed service execution environment based on the DINA active platform. We also described the context delivery system that is integrated into the DINA nodes and provides the required context for the different service components. In the next chapter, we demonstrate the advantages of the CONTEXT system by describing different service scenarios, and indicating the ways in which the CONTEXT infrastructure can be used to provide fast delivery of various services in a modern heterogenous networking environment.

References

1. Kornblum J, Raz D, Shavitt Y. 'The Active Process Interaction with Its Environment.' IWAN 2000, October 2000.
2. Housley R, Ford W, Polk W, Solo D. Internet X.509 Public Key 'Infrastructure Certificate and CRL Profile,' RFC 2459, 1999.
3. Adams C, Farrell S, Mononen T. 'Internet X.509 Public Key Infrastructure Certificate Management Protocol (CMP),' RFC 4210, 2005.
4. Alexander DS, Braden B, Gunter CA, Jackson WA, Keromytis AD, Milden GA, Wetherall DA. 'Active Network Encapsulation Protocol (ANEP).' Active Networks Group Draft, July 1997.
5. Deering S, Hinden R. 'Internet Protocol, Version 6 (IPv6) Specification,' RFC 2460, 1998.

8

System Evaluation

This chapter contains an evaluation of the system described in the previous two chapters. As explained in great detail in these chapters, the CONTEXT system provides an infrastructure for the fast development and deployment of efficient context-aware services. Thus, the first step in evaluating such a system is to show that context-aware services can indeed be developed and deployed using the provided infrastructure. In order to do so we introduce in this chapter several scenarios, and in each scenario we describe context-aware services built using the CONTEXT system. This demonstrates the ability of the proposed system to support the new and different types of service needed in today's telecommunications market. In the second part of the chapter, which addresses the efficiency of the system, we describe several benchmark measurements that test the scalability and efficiency of a single DINA element to support concurrent applications.

8.1. The Scenarios

In this section we describe the three scenarios, each addressing a different area in which context-aware services can be used. We describe example services that are useful in these scenarios, and how the CONTEXT system infrastructure is used to create, deliver, and control these services.

8.1.1. Work From Anywhere (WFA)

In the WFA service, a company subscribes to a Work Form Anywhere (WFA) service for its employees to access data on the move. This service bundles GPRS and a WLAN service, and promises a secure usable wireless data connection almost anywhere in the country with a high-speed connection at its many WiFi hotspots. These hotspots are located in many public areas including airports, railway stations, and public buildings such as hospitals, coffee shops, and the central business

Fast and Efficient Context-Aware Services   Danny Raz, Arto Tapani Juhola,
Joan Serrat-Fernandez, Alex Galis   © 2006 John Wiley & Sons, Ltd

districts. While access through these networks is available without the need for the CONTEXT system, the service described here allows subscribers transparent and flexible handover between a GPRS and WLAN. In addition, the policy-based management system coordinates the smooth creation, deployment, and execution of the service.

A typical scenario in which such a service will be used by an end user is as follows. *Consider Katherine, a graphic designer with three children. Katherine works from home a few days a week, using her home network to connect to the office network. On the scheduled day for completion of the project, she has been working on for the last few weeks, Peter, her 9-year-old son, complains that he does not feel well. His condition quickly worsens to the point that she has to take him to the local hospital. She calls a taxi to take them to the hospital and attempts to finish her project on the way. Halfway to the hospital she finishes and decides to send her work to the office using the GPRS network. However, the bandwidth of the mobile network is insufficient and it would take about 50 hours for the transfer of the 1 Gbyte file. In spite of this, she is not worried as her company had recently signed up to the Work From Anywhere (WFA) Data Service. This service provides an always-on low-bandwidth GPRS data transfer service and a very high-speed service at WiFi hotspots. Katherine knows there are several of these hotspots in the town, in particular at the St. Jude's Children's Hospital where she is taking Peter. When she arrives at the hospital, a handover from GPRS occurs and now the file upload will be complete in less than 2 hours. Katherine looks at the new transfer time estimate, lets out a sigh of relief, and holds Peter closer.*

In this subsection we describe the CONTEXT approach to the creation of the WFA service, as described in the previous two chapters. We also describe the building blocks from which the service is constructed.

8.1.1.1. Context Information

Four categories of context are relevant to this type of service: user context, network context, application context and location context. The specification of the service that is available to the user depends on all these aspects of context.

- *User Context* This relates to information about the user of the Context-Aware Service, that is Katherine in our described scenario. This includes her name, company, address, preferences, and profile. Other parameters such as her job title, skills, responsibilities, agenda, and tasks may also be included. The type of service that she has subscribed to, along with her user certificate, passwords, and service usage summary are also relevant.
- *Location Context* This information relates to the location of the user. It may specify whether the user is at home, in her office, in the hospital, or on the road (coverage area). It may also contain information regarding the type of network

access available at this location — a GPRS coverage area or a WiFi hotspot — and specific information about the access point or operator base station that covers the user's location.

- *Network Context* This is the relevant information about the networks that are available in the user's location. This information is constructed from static information about the underlying network topology and bandwidth, and from dynamic information obtained by monitoring the network, for example configuration parameters, bandwidth availability, latency, current NPN endpoints, network addresses, traffic levels, routing information, and network security. Network context encompasses information about the available access networks, and also network-related user information such as laptop MAC addresses, the access card type, and the assigned IP address.
- *Application Context* This context relates to the applications that the end user is using, whether she is using HTTP, FTP, or another protocol, whether she is using a VPN, the amount of traffic produced, and bandwidth consumption of the applications and the amount of time that the resources are being used. This information can be obtained by an agent running on the user machine, or (in a more complicated and less scalable way) by monitoring the user traffic and analyzing it.

In our scenario Katherine's context may be as follows. While within range of the hospital WiFi hotspot, the service will allow Katherine to have a secure (via VPN), high-bandwidth (the exact amount depending on network conditions) connection to her office. The application context will indicate that she is uploading a file-using FTP and so the service will be optimized for this type of traffic. Recall that in the CONTEXT system the context information is made available to the service logic through the context broker; thus, we need to describe how this information is collected and used in order to provide the service.

8.1.1.2. A Simplified Network Set Up

In order to describe deployment of the WFA service, we consider a simplified network topology and configuration in order to explain the network location of the different components of the system. Figure 8.1 depicts the main components needed to demonstrate this service: A GPRS domain, a WLAN domain, and a router with VPN capabilities representing the company domain. These domains are linked via the Internet with the Context-Aware Service Provider network monitoring the overall system. In Figure 8.1, the active routers host the DINA platform. The routers closest to the GPRS and WLAN also host a wrapper for the appropriate access network. The laptop is the user's mobile terminal, which should be connected to the company server, and sends and receives data. It contains a card that enables both GPRS and WiFi access. It is the CONTEXT view that one should not develop a heavy client,

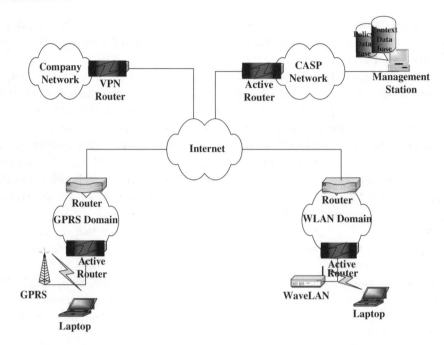

Figure 8.1   A Simplified Network Architecture for the WFA Service.

and so the laptop software and configuration should be changed as little as possible. The management entity resides in the Context-Aware Service Provider domain and contains the policy and context database, the AAA repository and the management applications for the entire system. It is assumed that mobile IP is supported by this network and any required redirection agents are located at the relevant routers.

This network will be used to demonstrate how the CONTEXT system works with respect to the service life cycle and especially when a user is using the service while changing location and entering a new WLAN domain.

8.1.1.3. Specification of a CAS and Management Policies

As described in the previous two chapters, when a new CAS is created, the CAS code generator generates the subscriber's service code and policies for that service. The code and policies are customized for every user and distributed to code storage points by the code distributor action consumer. Condition Evaluators (CEs) are installed and when the context changes (e.g., in the WFA case, when the user enters a new WLAN domain) they notify the policy decision-making component. As a result, according to the service definition, the relevant portion of the code is downloaded from the storage points into the code execution points by the code execution controller action consumer (See Chapter 6), and the code starts to run in the appropriate DINA node. At this point the service is invoked and the service assurance for this service is activated.

We concentrate on the service invocation and the code execution with respect to the WFA service. When a subscriber actually asks to use the WFA service, an event is raised and the service code is downloaded and invoked on the appropriate execution points, that is access points near the user (the user location is part of the context here). When the user logs out of the service, the service is suspended by stopping the code execution. During the time that the service is active, several events may occur. In order to show how the CONTEXT system actually provides a smooth transparent transaction between different access networks, a more detailed technical description is required. The following list summarizes the major events, conditions, and actions for this service. For each event, the parameters made available and the components that generate this information are specified.

Events

1. Mobile_User_WiFi_Event: This event occurs when a new incoming mobile user appears in a wireless coverage area.

- Event Variables
  - User_Id: This is the identifier of the user.
  - Password: This is the unique password to access the ASW service.
  - User_MAC_Address: This parameter contains the MAC address of the wireless card or device used by the user who is going to connect to the network. This parameter is used mainly for identification purposes.
  - Location: This variable states the location of the user (user's laptop or networked device). This variable refers to the node attached to the access point, which the user is connected to.

- Monitoring Component: Service Invocation Condition Evaluator — this is the CE that is executed when a user activates the service.

2. Mobile_User_Out_WiFi_Event: This event occurs when a mobile user sends a logout request to stop using the service. The logout request contains information identifying the service that the user wants to stop — other details are omitted in this case.

Conditions

1. Invocation of the WFA Service: This condition establishes how and when the management system decides that a user is demanding the execution of the WFA service.

2. Logout of the WFA Service: This condition establishes how and when the management system decides that a user wants to stop the service execution.

Actions

1. Execute_Service_Code: This action installs and executes the WFA Service Code (SLOs) on the appropriate Execution Points (i.e., DINA nodes)

   • Inputs

     – Service_Id: This is the identifier of the service (WFA), to be executed.
     – Subscription_Id: This is a unique identifier of the particular subscription to the service.
     – Customization_Id: This parameter identifies the particular customized code of the same service and subscription that needs to be executed.
     – Execution_Point_Required_Resources: This parameter specifies the resources that the execution point must possess in order to execute this particular service code. These resources may include minimum required free memory and CPU.
     – Execution_Points_List: This is a list of the execution points at which the service code must be started after its invocation.
     – Storage_Point_Selection_Criteria: This parameter specifies the criteria to select the storage point from where the service code will be downloaded into the selected execution points.
     – Customised_Runtime_Parameters: This parameter contains a list of runtime parameters specific to the service code to be executed. These parameters will be used to start the execution of the code.

   • Output

     – Action_Result: This variable contains a value indicating whether or not the action has been successfully executed.
     – Start Time: This variable states the time at which the user service code is started.

   • Enforcement Component: Code Execution Controller Action Consumer.

2. Stop_Service_Code: The service code is stopped when the user logs out. The Code Execution Controller Action Consumer can obtain the execution points where the code is running from the data tables. Details are omitted in this case.

Condition – Action Association

1. Service_Invocation_Policy: When the Invocation for WFA service condition is fulfilled, the Execute_Service_Code action is executed.

2. Stop_Service _Policy: When the Logout of the WFA service condition is fulfilled, the Stop_Service_Code action is executed.

8.1.1.4. Service Assurance

The PBSM is responsible for assuring that the service remains available. One key indicator of service availability is the continued execution of the service code. This is verified by receipt of a periodic 'I am up' signal from the service code. If an 'I am up' signal is not received in a specific time period, the PBSM will assume that the service is not running and it will attempt to restart the code.

Variables

1. Current_Timeout_Event: This variable states the current time and is given in 24-hour time format.

   - Monitoring Component: Service Assurance Condition Evaluator.

2. Time_Of_Last_Iamup_Signal: This variable states the time at which the last 'I am up' signal, from the SLO in question, was generated. The SLO periodically sends an 'I am up' signal. If it does, it means it is still alive.

   - Monitoring Component: Service Assurance Condition Evaluator,
   - Monitoring Parameter: Code_Id, the identification of the particular SLO to be monitored.

3. Time_Since_Last_Iamup: This variable states the time (in seconds) since the last 'I am up' signal was sent by the SLO.

Condition

1. Service_Code_Timeout: An 'I am up' signal has not been received from a service execution point within the defined Timeout_Interval. The condition is fulfilled when Time_Since_Last_Iamup_Signal > Timeout_Interval.

   When this condition is fulfilled, a Service_Timeout_Event is raised and a Service_Code_Restart action is executed.

Action

1. Service_Code_Restart: The service code is restarted if the PBSM realizes that it has stopped running but should be running.

   - Inputs
     - Service_Id: This is the identifier of the service whose code will be restarted.

- Subscription_Id: Unique identifier of the subscription to the service whose code will be restarted.
- Customisation_Id: This parameter identifies the particular customized code of the same service and subscription to be executed.
- Execution_Point_Required_Resources: This parameter specifies the resources that the execution point must possess in order to execute this particular service code. These resources may include minimum required free memory and CPU.
- Execution_Points_List: This is a list of the execution points at which the service code must be started after its invocation.
- Execution_Point: This parameter identifies the execution point at which the service code must be restarted to continue providing the service. This is the execution point where a service code crash has been detected.

• Outputs

- Action_Result: This parameter contains a value indicating whether the action has been successfully executed or not.
- Restart Time: This variable states the time at which the user service code is restarted.

• Enforcement Component: Code Execution Action Consumer

As described in the previous chapters, in order to deliver any service in the CONTEXT system one has to first use the authoring tools and define all the required elements and policies, and then define the logic of the service. We concentrate on three components of the code: the WFA SLO, which is the logic controlling the establishment of a QoS enable secure IP tunnel between a WiFi access point and the company network; the Service Invocation Condition Evaluator, responsible for detecting the entrance of a user to the WiFI coverage area; and the Service Assurance Condition Evaluator responsible for making sure the service is on and running. All these components are created during the CAS customization and personalization phases (see Chapter 6), and these components are attached to the personalized information of the specific user (Katherine).

The customized code components enter the management system and will be distributed according to the current policies. However, the code is not executed at this time. The execution policies will be triggered by events as described below. Note that, for scalability reasons, it may be better to have a general component that can be used throughout the customer organization rather than having a specific component for each user. However, in order to explain the sequence of events, we will consider an end-user personalization. The user location is monitored and made available as a context item. Based on information provided in the customization phase, all WiFi access points within a given distance of the user location are set to a 'Standby' mode.

In this case, a copy of the personalized Service Invocation Condition Evaluator is executed in all DINA nodes controlling the 'Standby' WiFi access points. Once the user is within range of an access point, a small application running in the user's machine (installed there as part of the service personalization) sends a message that is interpreted as context information. Using this information the system (as defined by the policy) decides to execute the SLO code in this DINA node. Once this is complete, the SLO uses the action broker and the QoS broker (as described in the previous chapter) to establish a secure IP tunnel providing the expected QoS to the end user. When the user moves out of range, the same small application sends a 'logoff' message and the SLO execution is terminated.

Of course, the real use of this service in a real heterogonous environment is more complex, and may involve many more components. However, it can be seen that once the CONTEXT infrastructure is available, self-adjusting context-aware services can easily be created and deployed in the network, providing a cost-effective way to provide such services to end clients.

8.1.2. Crisis-Aware Telecommunications Services

This is an example of a completely different type of service. In this case the user of the service is a telecommunications operator, not an end user as we saw in the previous example. Our goal in presenting this example is to show that the CONTEXT infrastructure can also be used in this case.

The purpose of this service is to allow an operator to change the behavior of its network during extreme emergency situations, such as a terrorist attack or major weather disaster. In such cases the available network resources should be managed differently from the way they are managed during normal day-to-day operation. For example, in order to ensure an increased number of users are able to access vital information, one might suppress nonvital traffic such as noncritical data transfer, or limit the duration of nonpriority calls in order to allow new emergency calls to be set up. Using the CONTEXT infrastructure, the network, which is managed using a policy-based approach, can self-detect emergency situations (e.g., by detecting a large number of calls to the emergency services) and automatically reconfigure the appropriate switches accordingly.

A typical scenario that demonstrates the use of such a service follows:

A plane crashes in suburb of a small town. Michelle, a frequent jogger, happened to be in the wrong place at the wrong time. Unfortunately, Michelle is now seriously injured by burning jet fuel. Mark, another jogger, who is fortunately carrying a cell phone, happens to arrive at the accident scene and makes a call to the emergency center. Since Mark is extremely upset, he cannot provide his location to the operator. In the mean time Jake, an eye-witness of the crash, calls a local newspaper in order to receive a reward for reporting the breaking news. There are hundreds of eye-witnesses who may be doing the same as Jake, or simply trying to contact their families and friends in order to tell them what they have just seen.

As unfortunately demonstrated in several terrorist events in recent years, many existing telecommunications networks are unable to cope with the very high call volume during such an event. However, if the switches in the area could be reconfigured, the network could be used to continue providing services to important callers during this time. For example, more resources could be directed to voice services, emergency workers could be given priority over 'normal' calls, and one could even consider limiting call duration in order to allow more users to call their loved ones.

Note that it is not clear that modern networks, which combine voice and data, should provide voice-only services in a crisis time. For example, maps of buildings could be essential to fire fighters, video conferencing may help paramedics receive medical help, and the distribution of suspect pictures may be very important to law enforcement agents during a crisis.

In order to provide such a crisis-aware service, one needs first to identify the crisis situation as it occurs, and then to enforce new resource utilization policies. In our scenario, the operator policy is to allow only emergency calls during a crisis situation, while other, lower priority calls are denied. In order to do this, one must identify if a call is an emergency call or not. One way to do this is by classifying phone numbers and comparing them to pre-defined lists.

Consider the arrival of ambulance with paramedic personnel Jane and Bob who start to take care of victims. Michelle is very seriously burned and Jane videophones the burns specialist for advice.

As the videophone consultation demands very high resolution, the network (controlled by the pre-defined policies) must limit the bandwidth of the other nonprivileged users. As described above, this is handled by special active services that manage network resources. The required configuration changes are automatically determined using context information such as the volume of the calls to emergency numbers, geographic origin of calls, etc., as described below.

The telecommunications network we are dealing with could be cellular, wired PTN, or VoIP. For simplicity of the technical explanation, we assume here that the network provides voice over IP using the SIP protocol. The Crisis-Aware Service can then identify a crisis situation by monitoring the call rate in each of the SIP softswitches, and when this rate increases suddenly, and more calls are made to the emergency service (112) it can declare an emergency and take appropriate action.

8.1.2.1. Realizing CATS in CONTEXT

Once the service is established, (there is no customization in this case) a monitoring SIP-SICE (Service Invocation Condition Evaluator) is distributed and installed in all DINA nodes responsible for SIP switches. The SICE uses the SIP broker in order to consistently monitor the amount of traffic and the number of emergency calls through the switch. Once a crisis situation is detected, an appropriate event is reported and a new configuration is uploaded to the switch. A new module containing the service logic is distributed and executed. This logic may change the Authentication Authorization and Access control (AAA) mechanisms of the switch. During regular operation the AAA mechanism is mainly responsible for making sure that calls are being made by authorized elements, and are charged according to the user plan. However in an emergency situation, resources (call establishment for example) are provided to emergency workers only.

We describe below the main parameters and sequence of events demonstrating the advantages of the CONTEXT system to handle these types of services.

8.1.2.2. CONTEXT Information

The context information used by this service is of the following nature:
Person Entity: User-related context: privileges of caller and called.
Network-Related Context: Number of sessions ongoing in the SIP softswitch whose destination is the 112 emergency telephone number at given time intervals.

8.1.2.3. SICEs, SLOs, and CCOs

The Crisis-Aware Telecommunications service scenario consists of two main components: the CH-SICE which is responsible for detecting emergency situations and triggering the change of the SIP softswitch control, and CH_Main which is the service code that determines the new AAA behavior under emergency conditions. CH_SICE polls a SIP broker (see Chapter 7) for sessions terminated at the 112 emergency number (or whatever the emergency number for that specific location is). This information is compiled into statistics. It can also be published as a context item using the Context Broker (as seen it Chapter 7). In this case CH_SICE becomes a CCO and the information (context) can be used by other services. Every DINA node that supports a SIP broker should have a copy of the CH_SICE so that the local SIP softswitch can be monitored. Note that the SIP broker might control several SIP softswitches, and that the CH_SICE monitors each one of them separately. Once the rate of calls to the emergency numbers exceeds a certain pre-defined threshold, CH-SICE determines that a crisis is occurring. The logic behind this is that usually each switch covers a geographic area, and a high rate of emergency calls from this area indicates an emergency event.

CH_Main is an active application that modifies the softswitch's normal behavior, and controls the SIP switch via the SIP broker. The service logic ensures that only critical calls are allowed in a crisis situation, while noncritical calls (or data services) are blocked or at least their allocated bandwidth is reduced. CH_Main runs on the DINA node that runs the SIP broker controlling the SIP softswitch located at the crisis area. Its execution is triggered by an event generated by the CH_SICE as explained above. This event is sent to the management system that identifies the control logic to be executed and the code repository to use, according to the pre-defined policies. CH_Main temporarily replaces the admission control function that is usually used by the SIP softswitch. It uses the SIP broker to install a redirection on the SIP node so that the SIP broker is invoked every time a new session is requested. In turn, the SIP broker redirects the request to CH_Main which decides whether the call should be admitted or not, based on the call parameters and the context information that details the roles of the different callers (defined by their telephone numbers).

8.1.2.4. Sequence of Interactions

A typical sequence of interactions per crisis is shown in Figure 8.2.

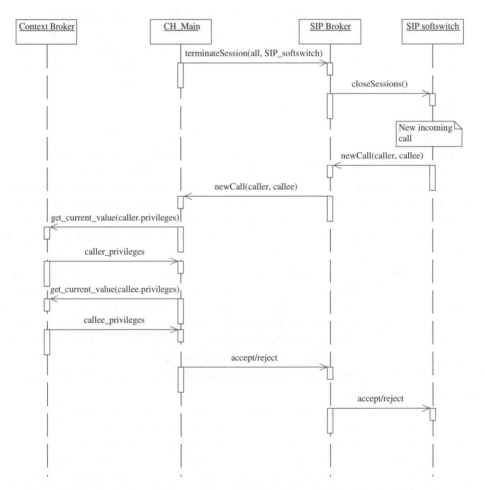

Figure 8.2   Sequence of Interactions of Crisis-Aware Service During Execution.

As explained above, the service starts when the PBMS receives an execution event raised by an SIP_SICE. Recall that this event is launched when the rate of 112 calls on any monitored softswitch exceeds a given threshold.

Once execution of CH_Main starts, all current sessions in the softswitch should be terminated, and all new sessions to be established will invoke CH_Main (the SIP softswitch informs the SIP broker which in turn informs CH_Main). CH_Main then uses the context broker to acquire information about the called and calling parties. Based on this information, CH_Main determines whether the call should be admitted. One can think of different ways to make the caller information available to the service logic; in this example we assume that the information is provided via the context system. In this way, this information can be used by this service and possibly by many other services as well.

As described in the previous chapters, in order to deliver this service in the CONTEXT system, one has to first use the authoring tools and define all the required elements and policies, and then define the logic of the service (i.e., CH_Main and the SICE). The code distribution policies (as described in Chapter 6) for each of the code components are similar: they both need level 1 and their potential execution point wildcard will refer to every DINA node that holds a SIP broker. Service assurance policies are designed to take care of possible failure of the service, and to make sure that the service is always available.

During the CAS customization phase (see Chapter 6) the subscriber is asked for several parameters, such as the reporting period for statistics (e.g., 5 minutes, 1 hour), the critical value to detect crisis, the emergency number (112 for Europe, 911 for USA), the areas to be covered, the filters to be applied, etc. The result is customized service logic – CH_Main and CH_SICE code, and the parameter values.

The customized CH_Main code is passed to the management system and will be distributed according to the current policies. However, the code is not executed at this time. The execution policies will be triggered by an event coming from CH_SICE. SIP_SICEs will be configured by the PBMS (as indicated above) to specific policies, and will use the SIP broker to monitor the wanted emergency call rate.

In an emergency situation (or actually when the emergency call rate exceeds the pre-defined threshold) CH_SICE will be triggered by an alarm. It then creates a new event containing the required information (SIP switch location) and sends the event to the management system.

The management system receives this event and matches it with currently installed policies. If CH_Main execution requirements are satisfied, its action part will be enforced. The first action will be to stop sending more events for the same area, so a message will be sent to CH_SICE. After receiving it, CH_SICE will stop monitoring the SIP switch in order to avoid creating more than one CH_Main instance for each switch in crisis.

The second action is to launch the CH_Main SLO to the switch in crisis. The code execution controller action consumer (see Chapter 6) asks the Code Distribution

Action Consumer for the URLs of the CH_Main SLO that are optimal for the new execution at the target node. This list of URLs is then injected in the Active Layer and causes the CH_Main code to be executed at the crisis DINA node.

As described earlier, CH_Main used the SIP broker to take over the switch access control, and uses the service logic to determine which calls are allowed and which should be blocked.

For example, all new calls to 112 will be blocked, because the first aid head-quarters have been already informed. Calls from an emergency team member (fireman, policeman, doctor) will be allowed. Calls to an emergency team member but originated by a nonemergency person (newspapers, even his family) will be blocked in order to avoid distracting them.

CH_Main will be stopped by a manual event raised by the operator when the crisis is over. This event will fire a reconfiguration action on CH_SICE to restart monitoring the SIP softswitch. At any time, assurance policies will ensure that CH_Main is not terminated unexpectedly and will restart the service if necessary.

The above description show how the CONTEXT infrastructure can be used by telecommunications operators in order to introduce, self-adaptive services, in an efficient, cost-effective way.

8.1.3. Moving Campus Services

In this scenario we consider the various modern multimedia services needed in a campus environment, and the way they can be realized using the CONTEXT system. In the university campus, the need for organizing videoconference meetings for exchanging information quite often arises. The *CA-Conference Set Up* Service is a context-aware service that establishes connections with QoS guarantees supporting streaming applications for conferences (videoconference, VoIP, etc), while taking into account the mobility of the participants, the capabilities of the access networks, as well as user-related information.

Consider the following scenario: *Professor John wishes to arrange a videoconference with professor Bob and the Research Associates, Alice and Nina at 18:00 in order to discus an ongoing project. In order to organize the videoconference, John utilizes the CA-Conference Setup Service. He accesses the relevant Web page and specifies the date, the start time, the duration, and the participants of the conference. The participants are accordingly notified and agree to participate. At 17:59, the CA-Conference service detects that the specified time for the conference is approaching. It then confirms that each participant can attend the conference and it searches the access network for each of the participants (John is connected to the network via Ethernet, Alice is connected to WLAN-AP1, John and Nina to WLAN-AP2, Bob is away from campus, but has a GPRS connection), and detects their terminal types and the type of application to be used for the conference.*

Considering this information, as well as the specified rules for network bandwidth allocation, CA-Conference service decides the bandwidth to be provided for each participant and sets up the relevant connections with QoS guarantees, and the conference finally takes place. During the conference, Nina moves to the upper floor of the building and as a result she automatically connects to the network via the WLAN-AP3. Consequently, the connection with QoS of Nina via WLAN Access Point B is removed and the rules for the new connection via WLAN-AP3 are set. However, it is also detected that the traffic to WLAN-AP1 is quite high and as a result Alice cannot properly attend the conference. So, Alice is advised to connect to the network through the WLAN-AP2. Furthermore, in case the signal of the proposed WLAN-AP is too low, the user is advised to move to another room, in order to achieve a better signal quality.

Even while participating in the conference, the involved parties are able to receive asynchronous announcements of other events of interest. The *CA-Announcement Service* provides this functionality. Provided that the participants have registered for this service and a relevant event is announced, the announcement will be delivered to the mobile phone or other terminal device carried by the participant, if it is inferred that the event fits his/her interests and preferences.

While participating in the conference, Alice who is a Research Associate interested in Wireless Communications, receives a notification that informs her about a lecture, on the subject of Wireless Sensor Networks, that will be given by a visiting Professor in 1 hour. The notification will reach her promptly as the service has inferred that Alice is not 'busy' and thus can be interrupted in order to be notified. On the other hand, Nina, who is a Research Associate interested in Pervasive Computing Issues, will receive the notification after the end of the conference, since the system has inferred that she is currently 'busy' and will be upset if it interrupted. Nina can still attend the lecture since the conference has finished just 10 minutes before the starting time of the lecture. If the lecture had finished before the end of the conference Nina would not be notified, because she should not be bothered with outdated notifications.

8.1.3.1. Mapping of Components to Scenario

The following figure depicts the actors involved in the scenario. Anyone can be an end user, provided he/she has access to at least one appropriate terminal and has subscribed to at least one service. Usually, end users own or have access to various types of terminals, such as mobile phones, PDAs, or PCs, and depend on one or more network providers for network connectivity. In this scenario we assume, for simplicity, a WLAN and fixed infrastructure provided as part of the campus network and a GPRS service provided by one of the major country providers. Stores and other facilities, such as the campus library, may want to offer context-aware services to end users, and for that purpose they will become service providers. When service

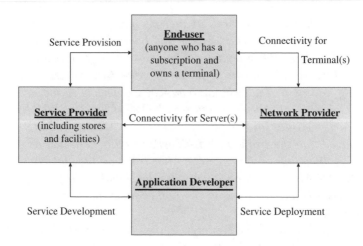

Figure 8.3   Interaction Between Actors.

providers have interesting ideas for services, they need application developers to write the source code and, when an application is ready, the service provider also needs a network provider, for the deployment of the final service (Figure 8.3).

In accordance with the service life cycle, the involvement of the Service Layer and Active Application Layer components described in the previous chapters is analyzed. During the creation phase, the Application Developer utilizes the functionality of *CAS Authoring* Component to produce the complete technical definition of each CAS involved in the Moving Campus Economy Scenario, namely *CA-Conference Set Up* Service and *CA-Announcement Service*. By selecting and combining the appropriate building blocks that represent the existing capabilities of the system, the application developer dictates the creation of the service definition documents. The selected capabilities depict the context information (e.g., the user's access network) and the existing services (e.g., the service that sets QoS) to be used by the each CAS.

In the following phase, the End User, through interaction with the *Service Customization* component, defines his personal details, such as profile and the provider of his agenda, and service customization details related to each CAS usage, such as the other participants, the starting time of the conference, the alternative devices/applications he wishes to use etc.

The service technical definition of each CAS, along with the produced customization, are fed to the *Code and Policies Generation Engine* component that is responsible for producing the working code for each customized service logic (SLO-Service Logic Class) and configuration policies. Subsequently, the SLO is stored and maintained at a selected point of the active network infrastructure. The operation of the SLO implementing the CA-Announcement Service is invoked as soon as the end user is detected in the Campus. However, the operation of the CA-Conference Set up is invoked as soon as the starting time of the specified conference is near (the functionality of the detection is provided by relevant Listeners). The relevant deployment and execution policies have been specified by the application developer during the creation phase and are utilized by the *Policy-based Service Management* components to deploy and execute each CAS.

During the Operation Phase, synchronous queries and asynchronous event notifications for context information are addressed to the local *Context Broker (CB)*. The communication between different CBs that are deployed on the distributed nodes of the network ensures that a requested context item, which is provided by a remote CB or a notification for a context event that occurs remotely, will reach the source CB. Moreover, each CB communicates with its registered CCOs for resolving complex context information. These CCOs are pre-defined and are deployed on the appropriate network nodes during the deployment of the SLO or in advance if they are shared by other operating CASs. Additionally, the existing services that implement the logic of the decisions taken by the SLO, such as setting QoS parameters in the access routers, are triggered through the interaction of the SLO with the *Action Broker*.

The *WLAN Broker* and the *GPRS Broker* provide information related to the mobile users connected to the current-access network. This information could be exploited in order to resolve queries about a user's type of network connection or IP address, and the presence or location of a user in the campus. Furthermore, information related to the characteristics of WLAN access points and GPRS network, such as the total network load of the WLAN access point, is also provided by WLAN Broker and GPRS Broker, respectively.

The action services in this scenario exploit the functionality of the *QoS* Broker for the DiffServ configuration of the domain's routers. The configuration of the core network can be done offline using the QoSBroker for each router. Once the required context is gathered, the QoS Broker sets up the relevant rules to the DiffServ domain's access routers. These rules refer to the installation of policer (traffic shaper/dropper) and marker modules for each user. The service's policy determines the parameters for these modules (e.g., the permitted transmission rate, the QoS level, etc). In this case, the implementation of a CCO is envisaged in order to calculate the bandwidth that should be allocated to each participant, taking into account several parameters. According to the logic of the SLO, a notification that is generated as a result of a user's movement will force the reconfiguration of the new access router.

The described scenario also requires the implementation of CCOs that control the generation of announcements and perform a match with the user's interests. Consequently, the user's profile should be accessed whenever a new announcement arises in order to send the relevant announcement notifications to a participant.

8.1.3.2. Identification of Context Information

The context-aware services that are part of the Moving Campus Economy scenario take into account a wide range of context data. This data, which affects the services' behavior, comprises of static as well as dynamic data and is acquired by various heterogeneous context sources. Additionally, the services require the acquisition of high-level context information that is produced by functions of interpretation, aggregation, or filtering implemented by CCOs. The types of context information to be considered are the following:

- Person Entity: User-related context: identity, details, interests/skills, preferences, and agenda.
- Place Entity: Location-related context: location information, user mobility information.
- Object Entity: Network-related context: type of access network, network capabilities/characteristics, available access networks, and capabilities/characteristics of them.
- Task Entity: Application-related context: type of applications installed on user's terminal and terminal-related context: type of user's terminal and capabilities.

8.1.3.3. CA-Conference Set Up

The CA-Conference Set up Service is responsible for providing QoS guarantees for a specific-time period, in order to hold a conference session between the members of a group. The organizer of the conference specifies the participants of the conference and its duration. The customized CA-Conference Set up SLO is executed as soon as the conference start time is reached. During the operation phase, the mobility of the participants is tracked and the appropriate QoS configurations are issued.

8.1.3.4. CA-Announcement Service

The CA-Announcement Service is responsible for delivering to the user asynchronous notifications/announcements about events of his/her interest. Once the user has registered for this service and an announcement for an event comes up, the notification is delivered to the mobile phone or the WLAN device a participant carries, as long as the event matches his/her interests.

8.1.4. Testbed and Service Layer Set Up

In order to describe the services and their implementation in the CONTEXT system, we use the simplified network depicted in Figure 8.4.

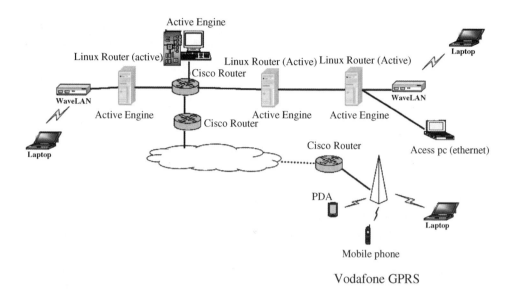

Figure 8.4   Simplified Network for the Moving Campus Economy Scenario.

The network consists of Linux and Cisco routers. The Linux routers are active, while each Cisco router is controlled by an active engine, which runs in a Linux PC. The DINA platform is installed on all Linux nodes. DINA active engines hosting WLAN Broker and GPRS Broker control the access routers of WLAN and GPRS network, respectively. The active engine that controls each router hosts a Context Broker, an Action Broker, and a QoS Broker. Moreover, the Mobility Broker is installed into all access routers.

Concerning the QoS issues, the network supports the Differentiated Services (DiffServ) architecture. Service classes (Expedited Forwarding, Assured Forwarding, and Best Effort) are installed into all the core routers of this simplified network. End-users can access the campus network via the following devices: personal computers, laptops, PDAs, and mobile phones. These devices may include Ethernet, WLAN, and GPRS cards for network access.

8.1.4.1. Description of the Sequence of Interactions

CA-Conference Set Up: The Conference Set up Service provides the QoS guarantees for a specific time period, in order to hold a conference session between the members of a group.

All the conference participants should be pre-registered to the service utilizing the Conference Set up Service web interface. When users register for this service, they must provide the following information. (a) Personal information (name, address, etc.) and (b) information about the network cards they own and its utilization abilities. MAC addresses for WLAN/LAN network cards and MSISDN numbers for GPRS cards, and (c) service level (users may choose among several service levels, which correspond to different accounting policies).

The conference session is scheduled by a registered user (service consumer) who utilizes the Conference Set up Service web interface in order to input the necessary information for the conference session. Specifically, he/she enters the conference start time, the conference duration, and the participants' names.

One way to realize this service in the CONTEXT system is by relating the definition of a conference to the customization phase as described in Chapter 6. When taking this approach, once the service consumer schedules a conference session, a customized SLO is created. Then, the SLO's source code is distributed and stored in the specified storage points. Moreover, an SICE is launched. The SICE produces a 'Start Time' event when the specified service execution time arrives. This event causes the SLO to be invoked.

The SLO logic asks to be notified when each of the participants is detected in the campus and has network access. Additionally, the access network and the end-user IP address are acquired. Then, the SLO triggers the relevant QoS configurations. If one of the connected users decides to change the access network, the SLO is accordingly notified and issues the new QoS setting to be configured. Moreover, if one of the connected users is accessing the network through a WLAN, the SLO asks to be notified in case it is possible to suggest a better network connection than the current one. In such a case, the SLO delivers the relevant information to the user. Finally, when the conference ends, the SLO triggers the removal of the performed QoS configurations.

The SLO interacts with the ContextBroker through the ContextBrokerInterface, in order to acquire the context information it requires, and with the Action Broker through the ActionBrokerInterface in order to trigger an action as described in Figure 8.5.

CA-Announcement Service: While participating in the conference, the involved parties are able to receive asynchronous notifications/announcements about events of interest. The CA-Announcement Service provides this functionality. Once users have registered for this service and an announcement for an event arises, the notification will be delivered to the participant's mobile phone or WLAN device, according to the relevant context information.

A user may register for this service utilizing the Announcement Service web interface. The information required is as follows: (a) personal information (name, address); (b) profile information identifying interesting activities; (c) details about the installed network card and its characteristics, (MAC addresses for WLAN network cards, MSISDN numbers for GPRS cards); and (d) the desired service duration (e.g., 6 months), which determines the user's accounting details.

Once a user registers for the service, a customized per-user SLO is created. Then, the corresponding source code is distributed and stored in the specified storage points.

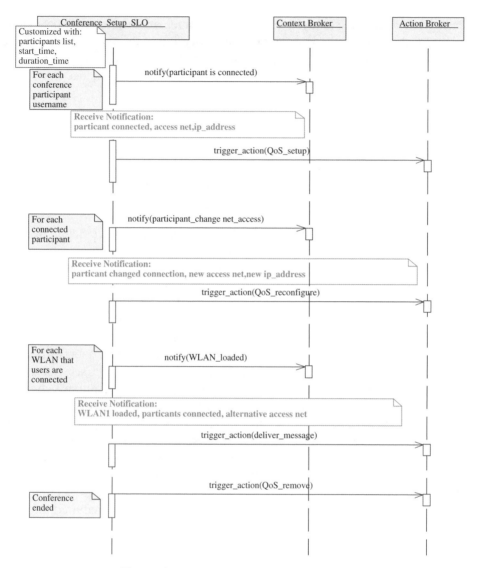

Figure 8.5   CA-conference Set Up Service.

Moreover, an SICE is launched and produces a 'User at campus' event when the registered user is detected in the Campus network. This event causes the SLO to be invoked.

As soon as a user is detected in the campus network, the SLO subscribes to be notified when an event matching the user's interests and preferences is announced. In this sense, it is required that once an announcement is produced, it is checked for a fit with the user's interests. In case of a match, the system consults the user's profile and agenda, while at the same time retrieving his/her location, in order to verify potentially pre-scheduled activities and proceed with the forwarding of the announcement to the current device, if his/her profile permits it. When the user leaves the campus, the SLO should be notified in order to stop checking for relevant announcements.

8.1.4.2. SICEs, CCOs and Actions

Along with the SLOs that implement the service logic, one has to identify a set of applications that comprise the SICEs, CCOs, and Action Applications.

The SICEs that should be implemented in order to realize the Moving Campus Scenario are the following:

- Start_Conference_SICE: This SICE is utilized for the Conference Set up Service. As soon as the SLO is created and stored by the Code Distribution component, the relevant policy triggers the configuration of the Start_Conference_SICE with the Start_Time of the scheduled conference and the conference id. The SICE accordingly produces the Start_Conference_event that invokes the execution of the customized Conference_Setup SLO through Code Execution Controller.
- User_Location_Monitoring_SICE: This SICE is utilized for the Announcement Service. As soon as the SLO is created and stored by the Code Distribution component, the relevant policy triggers the configuration of the User_Location_-Monitoring_SICE with the user's name and the monitoring location, which in our case is 'the campus.' The SICE is responsible for producing an event to signal the presence of the specified user in the campus grounds by detecting his/her local connection to the campus network. This invokes, through the PB-Service management system, the execution of the customized Announcement SLO. In order to produce the specified event, the SICE utilizes the ContextBrokerInterface for subscribing to receive through the Context Broker the required information from the network access points. Specifically, the SICE retrieves the required information through the Detect_WLAN_Users and Detect_GPRS_Users CCOs.

The CCOs are applications that produce complex context information used in the form of on-demand queries or events. They utilize the ContextBrokerInterface (as described in the previous chapter) to register the type of information that they provide ('register_new_context_object' method). After registering, they wait for context requests. The CCOs that are involved in our scenario are identified below:

- User_Info_CCO: This CCO is responsible for providing, upon request, the user subscription details such as the MAC addresses for WLAN/LAN network cards and MSISDN numbers for GPRS cards, or the service level of the user. Based on a given username, the CCO interacts with the storage place of this data and retrieves the relevant information.

– Detect_WLAN_Users CCO: This type of CCO is executed in each DINA active
engine that hosts a WLAN Broker (one per WLAN access point). The CCO is
responsible for verifying that a user is connected to the WLAN access point hosted
in the same node. Given a list of usernames (e.g. the conference participants
usernames), it interacts through the ContextBrokerInterface with the User_In-
fo_CCO and retrieves the MAC addresses of the WLAN network cards of the
users. Having these addresses, it can detect the users connected to the WLAN it is
attached to. The CCO queries the WLAN Broker utilizing the WLANBrokerInterface API (see Chapter 7), to find out if the participants are connected to the WLAN
network. Specifically, the CCO queries the WLAN Broker utilizing the 'getU-
serIP' function of the WLANBrokerInterface API. Moreover, a table containing
the (username-IP_ADDRESS) pairs for the connected users is created. Periodically, the CCO repeats the queries in order to detect possible changes (e.g., a user
disconnected from the WLAN access point or a new user has been detected). A
new table is created and it is compared to the old one, in order to detect users'
movement events. The relevant notification is delivered to the Context Broker,
which accordingly notifies the SLO that issued the request. An additional
parameter to be given to this CCO is the duration period of monitoring the user's
movement. This can be utilized for the case of the Conference_Setup Service, to
stop delivering notifications for the conference participants when the conference
session has ended.

– Detect_GPRS_Users CCO: This type of CCO is executed in the DINA active
engine that hosts the GPRS Wrapper. This CCO is responsible for providing
the information that a user is connected to GPRS. Given a list of usernames (the
participants usernames), it interacts through the ContextBrokerInterface with
the User_Info_CCO and retrieves the MSISDN addresses of the GPRS cards of
the users. The CCO queries the GPRS provider's MNO server utilizing the
GPRSWrapperInterface API to find out if the participants are connected to
the GPRS network. Specifically, the CCO queries the MNO server utilizing the
GPRSWrapperInterface API methods 'isUserConnected,' 'getUserIPAddress.'
Moreover, the CCO can retrieve the user's location in geographical coordinates
utilizing the GPRSWrapperInterface API method 'getUserLocation.' This
option is enabled in case the CCO is queried by the Context Broker on
behalf of the User_Location_Monitoring_SICE regarding the user's presence
in campus. When the user is connected, the relevant notification is delivered to
the Context Broker that accordingly forwards the notification to the SLO that
issued the request. Moreover, a table is created which contains the above
(username-IP_ADDRESS) pairs. Periodically, the CCO repeats the queries
in order to detect possible changes (e.g., a user disconnected from the GPRS
network or a new user has been detected), in which case a new table is created.
Regarding the monitoring time, the same as for the Detect_WLAN_Users
holds.

– WLAN_Monitoring CCO: In case one of the connected users is accessing the network through a WLAN, the SLO may ask to be notified when the system infers that the corresponding user would experience improved network quality by changing his (her) access point. Consequently, this CCO is used for the traffic monitoring of the WLAN access points. Specifically, it periodically retrieves the network load of the access points (for example, every 30 seconds) and compares them with predefined thresholds. If the network load of an access point exceeds the threshold, the application retrieves the IP addresses of the participants that are connected to this WLAN access point and calculates the users signal quality percentage. If an interested user has a relatively low signal quality, the WLAN_Monitoring CCO sends a message to him/her, suggesting a move to another WLAN access point. The message also details how to find the proposed access point (e.g., 'Go to the second floor'). The WLAN_Monitoring CCO utilizes the   following   WLANBrokerInterface   API methods: 'getLoad,' 'getSignalQuality'.

– Announcement_Monitoring CCO: This is a general CCO that is permanently active and monitors the sources of announcements. Its objective is to generate a notification when a new event is announced.

– User_Matching_Announcement CCO: This is a customized per user CCO that checks if an incoming announcement matches the user's profile. It takes the identifier of the announcement and the username as input, and decides whether the announcement could interest the user and should thus be delivered to him/her, based on the retrieved user's profile. The result is the generation of a message that indicates the result of the matching process.

The action applications to be triggered are the following:

- QoS_Setup: This action application is responsible for performing the necessary QoS configurations as specified by the Conference_Setup_SLO. When triggering this application the parameters conference_duration and RULES are also given. Conference_duration stands for the time period for the configurations, while RULES represents the configuration properties (participants' IP addresses, their gateways' IP addresses and participants' service level). An example of the RULES parameter is given below:

⟨RULES⟩
 ⟨SOURCES⟩
  ⟨SOURCE⟩
    ⟨IP_ADDRESS⟩147.102.7.45⟨/IP_ADDRESS⟩
    ⟨PORT⟩2345⟨/PORT⟩
    ⟨SERVICE_TYPE⟩EF⟨/SERVICE_TYPE⟩
    ⟨RATE⟩100⟨/RATE⟩
    ⟨BURST⟩200⟨/BURST⟩
      ⟨SERVICE_TYPE⟩AF11⟨/SERVICE_TYPE⟩

```
    ⟨RATE⟩50⟨/RATE⟩
  ⟨BURST⟩200⟨/BURST⟩
  ⟨SERVICE_TYPE⟩BE⟨/SERVICE_TYPE⟩
    ⟨RATE⟩50⟨/RATE⟩
  ⟨BURST⟩200⟨/BURST⟩
⟨/SOURCE⟩
⟨SOURCE⟩
    ⟨IP_ADDRESS⟩147.102.7.52⟨/IP_ADDRESS⟩
    ⟨SERVICE_TYPE⟩EF⟨/SERVICE_TYPE⟩
  ⟨PROTOCOL⟩UDP⟨/PROTOCOL⟩
    ⟨RATE⟩200⟨/RATE⟩
  ⟨/SOURCE⟩
  ⟨/SOURCES⟩
    ⟨DESTINATIONS⟩
  ⟨DESTINATION⟩
    ⟨IP_ADDRESS⟩147.102.7.45⟨/IP_ADDRESS⟩
    ⟨PORT⟩2455⟨/PORT⟩
    ⟨/DESTINATION⟩
  ⟨DESTINATION⟩
    ⟨IP_ADDRESS⟩147.102.7.45⟨/IP_ADDRESS⟩
    ⟨PORT⟩2455⟨/PORT⟩
    ⟨/DESTINATION⟩
  ⟨/DESTINATIONS⟩
⟨/RULES⟩
```

The application parses the RULES parameter to retrieve the necessary configuration actions. The QoS_Setup application communicates with the QoS Broker utilizing the QoSBrokerInterface API. It considers all the (source, destination) combinations in order to install the necessary rules. One TCP session with the QoS Broker is required for each (source, destination) pair. Finally, during the lifetime of the specified conference_duration the QoS_Setup application may be used for reconfiguring the QoS settings as due by the users' movement.

- send message: This action provides the functionality required to send a message to the appropriate user. Based on the access network, the IP address of the user and the message itself, it will construct and deliver the appropriate message to the user. This functionality is required for delivery of the message advising the user to move to an alternative access point if the current one is overloaded, as well for the delivery of all the announcements in the case of the CA-Announcement Service.

This very detailed description of a possible realization of the Moving Campus Services in the CONTEXT system indeed shows that how the design of the system can be used to generate scalable efficient CASs in this environment. Overall, the three scenarios described in this section indicate that the proposed system can be used in different ways to allow fast creation and deployment of efficient context-aware services in heterogeneous networks.

8.2. Performance Evaluation

In the first part of this chapter we evaluated the CONTEXT system by showing how various context-aware services can be deployed in the network using the CONTEXT infrastructure. A second part of this evaluation consists of assessing the scalability of such a system and the expected performance. As explained in the previous two chapters, the distributed heart of the system lies in the collection of DINA machines deployed in the network. These machines can be viewed a distributed execution environment where the service logic (and other components) are executed. Thus, a first step in evaluating the performance and scalability of the CONTEXT system is to examine the performance of a single DINA node when executing the logic of many services.

8.2.1. CPU Load

In order to check the performance in terms of CPU utilization, we designed a benchmark application that represents service logic that is bounded by computation resources. This application is sent and executed in a DINA node. Without any load the application requires about 2 seconds to complete. Then, we added further load by executing a number of load application representing the logic of other (different) services running on the same DINA node.

As one can see from Figure 8.6, when the load increases the time it takes the application to complete increases as well. The Total time in this figure is the time taken to load and execute the application, while the Execution time is the amount of

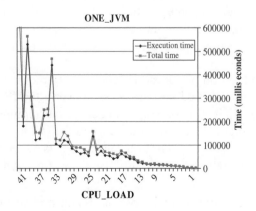

Figure 8.6   Load One JVM.

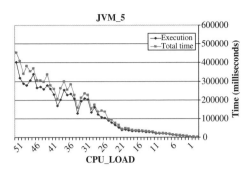

Figure 8.7   Load Five JVMs.

CPU time in the destination DINA node. Clearly, as we add more CPU intensive applications (i.e., increasing the load), the amount of resources available to the benchmark application reduces, and hence the amount of time required to finish the task increases. The almost linear increase indicates that the overhead of managing many applications at the same node is relatively small. However, at some point, around 30 CPU intensive applications, the system becomes unstable and the required time for termination increases sharply. This point indicates that the load had reached its critical point, and increasing the load above this point may cause undesired behavior such as timeouts and the inability to perform all services. Recall from Chapter 7 that in order to address scalability and load issues, the design of the DINA system allows several JVMs to be deployed on the same DINA node.

Figure 8.7 shows the completion time of our benchmark application when five JVMs where present in the DINA node. In such a case the CPU intensive load applications are shared among all JVMs almost equally; thus, the JVM in which the benchmark application is executed has only one fifth of the load it would have if only one JVM was used. However, the overall CPU usage is almost the same since we have five JVMs each having about one fifth of the load and one can see that the critical point appears at about the same load. Note that the system becomes much more stable when the number of JVMs increases and the variant of the execution time becomes much smaller when we move from one JVM to five JVMs and then to Ten JVMs in Figure 8.8. Clearly, if each JVM would have its own physical machine within close proximity of the DINA node, we could expect a dramatic improvement in performance, since in such a case we would have five (or ten) times more CPU resource available.

8.2.2. *Info-Broker Load*

In many applications, and in particular in the logic of services, CPU is not the main bottleneck, and most of the time the service is waiting for data. In order to test the

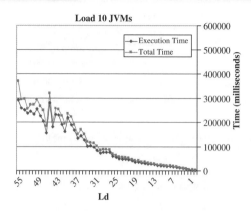

Figure 8.8   Load Ten JVMs.

performance of such services, we created a different benchmark application. This application uses the InfoBroker in the DINA node, representing a service logic application that needs access to local information. The load creating applications were also changed in a way that they generate load on the InfoBroker as well. As one can see in Figure 8.9, the time taken to complete the task increases as the number of load applications increases. Again at some point (around 50 load applications in our case) the system becomes less stable, but up to this point the execution time increases linearly with the number of load applications, and the InfoBroker seems to handle the load efficiently. It is predicted that other brokers will perform similarly, depending of course on the specific implementation of the broker.

The actual performance of the CONTEXT system in real scenarios depends on many parameters and can be tested only when the system is deployed in large-scale configurations (i.e., many representative applications executing in a realistic CONTEXT infrastructure). However, the preliminary results of the performance testing

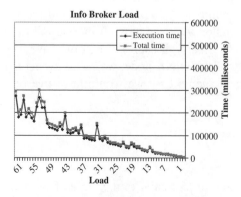

Figure 8.9   InfoBroker Load.

presented in this chapter indicates that it has a potential to be a very scalable system, providing the required infrastructure for the fast and easy deployment of efficient context-aware services in heterogonous networks.

8.3. Conclusions

The CONTEXT system provides an infrastructure for the fast development and deployment of efficient context-aware services. An evaluation of the CONTEXT system is presented in this chapter. The first step in evaluating such a system is to show that context-aware services can indeed be developed and deployed using the provided infrastructure. In order to do so we introduce several scenarios, and in each scenario we describe context-aware services built using the CONTEXT system. This demonstrates the ability of the proposed system to support the new and different types of service needed in today's telecommunications market. In the second part of the chapter, which addresses the efficiency of the system, we describe several benchmark measurements that test the scalability and efficiency of a single DINA element to support concurrent applications.

9

Conclusions

Next-generation networks are driven by the convergence of voice and data into fully integrated networks. Such converged networks will be characterized by the increasing number of wireless and cellular users that are always connected via WiFi (802.11), WIMAX (802.16), or various 3G and legacy cellular technologies, the move towards overlay networks and Peer to Peer (P2P) applications, and the deployment of both traditional telecommunications services and new data services that require QoS support.

Considering the small margins in the market and the increasing competition, many providers seek to offer new services that can both attract new costumers and become the source of substantial future revenue. A successful service, therefore, is one that in addition to offering new experiences to end-users is also profitable for the provider. From this perspective, two very important aspects of such sophisticated new services are the time to market and the operational cost. This market drive for new advanced services in the converged world of wireless voice and data brings us faster than ever to the time when Context-Aware Services (CASs) are a mainstream service and a major source of income for service providers.

9.1. Context-Aware Services

By their name (and definition see Reference [1]) CASs are services in which the actual result of using them depends on the context. The interest in such services started in the early 90s in the field of Pervasive Computing, where context was usually associated with the user location and information provided by various sensors (gadgets). In the wireless communication world, services that react according to the location of the user (and other users) answer questions such as 'where is the nearest Italian restaurant?' or 'who is next to me?' and are already being offered to cellular customers. The convergence of voice and data networks, and the rapid

Fast and Efficient Context-Aware Services   Danny Raz, Arto Tapani Juhola,
Joan Serrat-Fernandez, Alex Galis   © 2006 John Wiley & Sons, Ltd

growth of the wireless world create the need for more sophisticated CASs, in which the context is more than just the user location.

Although the concept (and name) of CASs has only been developed in the last decade (starting with the fundamental paper of Bill N. Schilit *et al.* [2]), Context-Aware Services have existed in the telecommunication world for a long time. Perhaps the most basic example is the emergency call mechanism. Emergency calls can be viewed as a service in which the user dials a fixed number (911 in the US and 112 in Europe) but the effect is different, depending on the location of the caller, and sometimes the status of the emergency call centers or the local police station. Based on these 'contexts' the call is redirected to the appropriate location. Other advanced voice services, such as follow me, can also be viewed in the same way where the result of dialling a number depends on the context, which is the current location of the recipient of the call (according to the information available to the network). In an enhancement of these services, in the spirit of Pervasive Computing, sensors detect the current location of a person in the building and forward his (her) calls to the phone located in the same room, as described in Reference [4]. The basic emergency call service can also be extended in various ways as described in Reference [3].

As for traditional data services, consider a very common service like web browsing, in which we would like pages to be loaded quickly, which might entail connecting to the closest replica of the pages we are looking for, or to the server with the lowest load or fastest response time. If we envision this as normal web browsing that redirects the page request according to the parameters (server load, traffic load) above, then this is also a Context-Aware Service, where the context consists of the client's network location, the location of the replicas in the network, the load on each replica server, the network traffic, and the current routing paths. Such a service is much more network oriented, but the main concepts of CAS are still valid.

A more complex example is the 'smart follow me email' system [5]. In such a system, the way e-mail is forwarded to the user depends on several parameters, for example, the type of device the user is using (PDA, cell phone, laptop, etc.), the type of connectivity and available bandwidth (GPRS, WIFI, modem, ADSL, etc.), and the importance of the information to the person at this particular time. For example, JPEG pictures are not forwarded to the user when using her PDA and a GPRS cellular data connection (due to the cost and large amount of bandwidth required), unless the e-mail contains the map with the driving instructions for the location of the next meeting. In this case the service reacts according to the context, which is composed of: user location, user private information (meeting schedule), e-mail content, connection type, and available bandwidth.

This last example demonstrates the difficulty of developing complex CASs in the converged world of voice and data. The context information needed by the service is complex; it comes from different sources and (at least some of it) is not managed by the service provider. Moreover, some of the information is network-context information, which should be collected from a distributed environment, and several of the technical aspects of providing such a service in a scalable efficient fashion require access to, and ability to configure, elements in the networks.

In this book we describe the state of the art in this field and a new framework aimed at addressing the need for rapid deployment of efficient Context-Aware Services, which is becoming a requirement in many providers' networks. This solution is based on a distributed service execution environment utilizing the programmable network paradigm.

A study of the field of Context-Aware Services followed development of new ways to create and deploy such services must start with a clear view of the different elements participating in the envisaged scenarios. A key ingredient is the context itself – one must define what context is, and how it relates to services in the new networking paradigm. This is described in Chapter 2 of this book. Another important building block is the service. Here, it is important to define the scope of a service in this new era where telecommunication and data networks converge. It is also important to study the service life cycle, from creation through deployment to the actual offering of well-managed services to end users. We do this in Chapter 3.

A special emphasis is put on the interaction of the service with the networking layer. In this new converged world where the network is very heterogeneous and complex, and where low operational cost is crucial to the attainment of profitability, it is very important to be able to offer the required QoS to the customer in the most efficient way. For this reason, it is no longer possible to offer all services from a single location and to view the network as a black box. Thus, there is a clear need for a well-defined control and management API between the services and the network elements. We discuss this important aspect in Chapter 4.

A good way to maximize the advantage of such an API, and to allow distributed applications to cooperate in offering the service, is to use programmability. Network programming techniques allow the creation of a distributed service execution environment that can host the service logic and utilize the service network API. This approach is followed in the development of the CONTEXT system that is described in this book. Programmable technology and its applicability to services are described in the Chapter 5.

The CONTEXT system is a middleware solution for efficient development and deployment of context-aware services making use of programmable system technology. This system consists of a distributed service execution environment (EE) composed of DINA nodes, and a service support layer (SSL) that is dedicated to the creation, customization, deployment, and management of services on top of the distributed EE. The details of the different layers are described in chapters 6 and 7 of the book.

In order to make sure that the proposed system can indeed be used by the different players in the service domain, it very important to examine the different ways the system could be used to create and deploy different types of service. In Chapter 8, we provide such an evaluation for describing different scenarios and discussing the ways the CONTEXT system is used in order to create, deploy, and manage these services. We also present evidence of the system's scalability by presenting key performance measurements taken on the system prototype.

It is important to note that the CONTEXT system (or similar common service infrastructure) is more than just an implementation technique that enables scalability and efficiency. In fact, once the common infrastructure is deployed by the service provider, creating a service will be generic in the sense that the same service can be developed once and deployed many times by many providers in different networks, possibly in different countries. This enables the creation of a new type of business – service developer. Such businesses can concentrate on market and client needs and develop corresponding novel services. The beauty of such a paradigm is that using the common infrastructure, these services can become off-the-shelf products, ready to be used by different ISPs all over the world.

9.2. Autonomic Communications Vision

Context awareness in networks and services is one of the key pre-requisites for realization of the *Autonomic Communications* vision. Autonomic communications systems are self-aware and they possess self-knowledge, continuously optimise and dynamically restructure themselves, adapt to (un)predictable conditions and changes to their environments, prevent and recover from failures, and provide a safe environment.

The key feature of autonomous communication systems is that they exhibit self-awareness capabilities, in particular self-contextualization and self-management.

Self-Contextualization – Contextualization is a communication service property. A context-aware system is able to use context information to improve the performance of its expected role, and also to maximize the perceived benefits of its use. Self-contextualization is the ability of a system to describe, use, and adapt its behavior to its own context. Once a service component becomes context aware, it can make use of context information for other self-management tasks that depend on context information. In this way context becomes a decisive factor in the success of future autonomous systems adaptive to changing conditions.

Self-Programmability – Programmable service networks take advantage of network processing resources by dynamically injecting new code into systems elements in order to create new functionality at run time. Applications and services are thus able to utilize required network support in terms of optimized network resources and as such they can be said to be network aware, that is a service-driven network. Self-programmability means that programmable service networks follow autonomous flows of control triggered and moderated by network events and changes in network context. The network is self-organized in the sense that it autonomically monitors available context in the network, and provides the required context (and any other necessary network service support to the requested services) and self-adapt when context changes.

Self-management – Currently, network management faces many challenges: complexity, data volume, data comprehension, changing rules, reactive monitoring, resource availability, and others. Self-management aims to automatically address these challenges through self-optimization, self-organization, self-configuration, and self-adaptation.

Self-optimisation – As network context information and resources, and their availability are changing rapidly, there is a need for an autonomous tool for consistent monitoring and control of network-context information and resources, so that service components may be executed or deployed in the most optimized fashion. Autonomic systems aim to improve their operational goals on a continuous basis. They must identify opportunities to make themselves more efficient from the point of view of strategic policies (performance, quality of operation, cost, quality of service, quality of context, etc.).

Self-organisation – This enables autonomous structuring of network-context information and resources, making them available to services. The autonomous structuring of network context information and resources is an essential self-organisation task. In order for services to make use of distributed context information and resources, networking elements will be (re)structured and referenced in an easy-to-access-and-retrieve structure in an automatic fashion. All network-context information and resources will be autonomously organized and reserved through a service layer.

Self-adaptation – Autonomic systems must configure and adapt themselves in accordance with high-level policies representing service agreements or business objectives, rules, events, and environments. When a component or a service is introduced, the system will incorporate it seamlessly and the rest of the system will adapt to its presence. In the case of components, they will register themselves and other components will be able to use them or modify their behavior to fit the new situation. Autonomic Systems satisfy the need for an open programmable self-configuring infrastructure.

Self-healing – Autonomic systems will detect, diagnose, and repair problems caused by network or system failures. Using knowledge about the system configuration, a problem-diagnosis embedded intelligence will analyze the monitored information. Then, the network will use its diagnostic functions to identify and enforce solutions, or alert a human in cases where no solutions can be found.

Self-protection – Autonomic systems will defend themselves as a whole or as components by reacting to, or actively anticipating, large-scale correlated problems arising from attacks or cascading failures that remain uncorrected by self-healing measures.

Undoubtedly, sophisticated context-aware services are going to take an important part in future converged telecommunication and data networks. This book describes the CONTEXT project view of a common infrastructure that supports scalable, efficient, and cost-effective services, a step on the path toward service-centric networks, built from full autonomic services.

References

1. Chen G, Kotz D. A survey of context-aware mobile computing research. Technical Report TR2000-381, Department of Computer Science, Dartmouth College, November 2000.
2. Schilit BN, Adams N, Want R. Context-aware computing applications. In *IEEE Workshop on Mobile Computing Systems and Applications*, Santa Cruz, CA, US, 1994.
3. Hegering HG, Kupper A. Management challenges of context-aware services in ubiquitous environments. Technical report, 2003.
4. Want R, Hopper A. Veronica Falčao, andJonathan Gibbons. The active badge location system. *ACM Transactions on Information Systems* 1992, **10**: 91–102.
5. Cohen R, Raz D. An open and modular approach for a context distribution system, IEEE/IFIP Network Operations and Management Symposium, NOMS 2004, April 2004, pp 365–379.
6. Kornblum J, Raz D, Shavitt Y. 'The active process interaction with its environment,' IWAN 2000, October 2000.

Index

Fast and Efficient Context-Aware Services Danny Raz, Arto Tapani Juhola,
Joan Serrat-Fernandez, Alex Galis © 2006 John Wiley & Sons, Ltd